Mechanics of Liquid Nano- and Microdispersed Magnetic Media

Mechanics of Liquid Nano- and Microdispersed Magnetic Media

V. M. Polunin
A. M. Storozhenko
P. A. Ryaplolv

CISP

CRC Press
Taylor & Francis Group
Boca Raton London New York

CRC Press is an imprint of the
Taylor & Francis Group, an **informa** business

CRC Press
Taylor & Francis Group
6000 Broken Sound Parkway NW, Suite 300
Boca Raton, FL 33487-2742

First issued in paperback 2020

© 2017 by CISP
CRC Press is an imprint of Taylor & Francis Group, an Informa business

No claim to original U.S. Government works

ISBN-13: 978-0-367-57321-8 (pbk)
ISBN-13: 978-1-138-06823-0 (hbk)

**Visit the Taylor & Francis Web site at
http://www.taylorandfrancis.com**

**and the CRC Press Web site at
http://www.crcpress.com**

Contents

Foreword

The textbook systematizes and describes a set of questions related to the mechanics of liquid nano- and microdispersed media, which have not been paid sufficient attention in the courses of general physics and which are described only in a few specialised publications and periodicals. At the same time, these issues directly correspond to the programs and tasks of nanotechnology and, therefore, the textbook of this direction is necessary and timely.

The difficulty of creating a textbook in which the material with this specificity is concentrated is, first of all, in the absence of the publication of a similar topic as a model for comparison. The authors of the textbook, proceeding from the volume of the course designed for one semester, made an attempt to select the material available in the domestic and world scientific literature that will allow describing the behaviour of these unique environments characterised by such 'incompatible' qualities as fluidity, compressibility and strong magnetism. In this textbook, on the basis of fundamental physical equations describing the mechanisms of interaction of liquid magnetic media with a magnetic field, we explain the physical nature of a number of specific effects, including the emergence of ponderomotive force, magnetic pressure jump, ponderomotive elasticity, the effect of 'slippage' of microparticles relative to the liquid matrix, Elastic vibrations and resonance in vibrational systems with a magneto-fluid inertial element, the effect of magnetic levitation, the magnetocaloric and magnetorheological effects. The influence of the particle size and polydispersity of the composition on the formation of the magnetization curve, on the phenomena of diffusion and magnetophoresis is discussed. The manual deals with the mechanisms of magnetisation of microdispersed suspensions and nanodispersed magnetic liquids, as well as an experimental technique and a description of the equipment for obtaining a magnetisation curve with subsequent use in magnetogranulometry. The methods for obtaining nanodispersed magnetic liquids and ferromagnetic suspensions are described, and the main existing and prospective applications of these media with the corresponding physical interpretation are listed. Examples are given of calculating the physical quantities, as well as control questions that can be used to consolidate the knowledge acquired in the process of learning and self-control. Realising that individual errors are possible in the first manual on this topic, the authors hope that it will help students to obtain the necessary physical representations in the field of mechanics of liquid nano- and microdispersed magnetic media.

The textbook is intended for bachelors, masters and students of full-time and part-time forms of teaching nanotechnology specialties. The textbook will also be useful for students studying at university courses in general physics, postgraduates of relevant physical specialties, young professionals and engineers.

Main notations and abbreviations

$\mathbf{B}$ — the magnetic induction vector

c — the propagation velocity of sound waves

C_p — specific heat at constant pressure

C_V — specific heat at constant volume

e — electromotive force

G — the gradient of the magnetic field strength

$\mathbf{H}$ — the vector of the magnetic field strength

k_0 — the Boltzmann constant

$L(\xi) = \text{cth}\,\xi - \xi^{-1}$ — Langevin function

$\xi = \mu_0 m_* H / k_0 T$ — the parameter in the Langevin function

$\mathbf{M}$ — the magnetization vector of matter

M_S — the saturation magnetization of a magnetic fluid

M_{S0} — the saturation magnetization of a ferroparticle

M_0 — the magnetisation of the medium in the unperturbed state

m_* — the magnetic moment of the particle

N — the demagnetising factor

n — the concentration of ferroparticles

p — pressure

Q — the quality factor

q — the coefficient of thermal expansion

R, r — is the radius, the coordinate of the cylindrical system

S — entropy, surface area, deformation

T — absolute temperature

t — time

V_f — the volume of the magnetic core of the particle

β_S — adiabatic compressibility

γ — the ratio of specific heats (adiabatic index or Poisson's ratio)

ε_0 — the electric constant

η — the viscosity coefficient, total viscosity

η_S — the shear viscosity coefficient

η_V — the bulk viscosity coefficient
μ_0 — the magnetic constant
ν — oscillation frequency
ρ — density
ρ_1, ρ_f —density of the carrier fluid
ρ_2, ρ_s —density of solid particles
τ_N — Néel relaxation time
τ_B — time of Brownian rotational motion of particles
φ — the volume concentration of the dispersed phase
χ — magnetic susceptibility, thermal conductivity coefficient
ω — circular oscillation frequency
k_g — the gas elasticity coefficient
k_σ — the coefficient of elasticity of surface tension
k_p — the coefficient of ponderomotive elasticity
λ — the wavelength
F, f — force
p_0 — atmospheric pressure
ρ_g — the density of gas (air)
σ — the surface tension coefficient, electrical conductivity
T_C — the Curie point

MF — magnetic fluid
MFS – magnetic fluid sealer
SAS – surfactant
FP — ferroparticle
FS — ferrosuspension

Introduction

The investigations of the physical properties of the microdispersed magnetised media started already in the first half of the 20th century with the investigation of the magnetic and rheological properties of fluids with ferromagnetic particles suspended in them with the size varying from several micrometres to tens of micrometres. These disperse systems are referred to as ferrosuspensions (FS). The practical application of these ferrosuspensions is based on the very strong dependence of viscosity on the strength of the magnetic field. The FS and paste-like compositions are used for the visualisation of the domain boundaries, in braking systems, in magnetic flaw inspection, in the manufacture of tape recorder tapes, in the technology of separation of iron or and in some other areas.

However, in the solution of a number of other practical problems, the anomalous magnetic dependence of viscosity is an interfering factor. Another shortcoming of the disperse systems of this type is their instability, the irreversible separation of magnetic and non-magnetic phases under the effect of the gravitational force or a heterogeneous magnetic field.

A qualitative 'jump' in the formation of stable fluid magnetically controlled media with a high stability of the structure and the almost complete independence of viscosity on the magnetic field was made in the 60s as a result of the development of magnetic fluids (MF) – one of the first products of the new technologies [1, 2].

The magnetic fluid is a colloidal solution of the single-domain ferri- and ferromagnetic particles in the carrier medium. To make sure that the disperse system has the required aggregate stability, the magnetic particles are coated with a monomolecular layer of a stabilising agent. In the final analysis, the solution of the problem of the development of a material not existing in nature with the required physical properties has been solved in the adjacent area of knowledge – the physics of ferromagnetism, colloidal chemistry and magnetic hydrodynamics. Because of the combination of the 'mutually excluding' property such as fluidity and the capacity to magnetise to saturation in relatively weak magnetic fields the micro- and nanodispersed magnetic media are used in various areas of science and technology: magnetic fluid seals (sealing agent), magnetically controlled lubrication in friction sections and supports, the separators of non-magnetic materials, the agents for cleaning the water surface to remove oil products, the sensors of the angle of inclination and acceleration, the filler of the gaps of magnetic heads of loudspeakers.

The fact that the information in the area of mechanics of fluid magnetised media, obtained by students in general university physics courses, is insufficient restricts their views regarding the possibilities of advanced technologies. On the other hand, the requirements of practical application put emphasis on the problems of including these problems in the educational process in order to prepare future graduates for participation in the development of new advanced devices and systems.

The main material the textbook is divided into 14 chapters. The equations of the dynamics of liquid magnetised media, including the ponderomotive force, are presented. Examples of manifestation of the ponderomotive force with a detailed analysis are described. The physical nature of the magnetic pressure jump at the interface of two magnetic disperse media is shown. A large number of examples are used to indicate the effect of the dimensional factor on the physical properties of the materials: the mechanics of 'slipping' of the nano- and microparticles in relation to the liquid matrix in the accelerated movement of the system is discussed; the rheological properties of the suspensions of colloids are studied; special features of magnetisation of microdispersedd suspensions and nanodispersedd magnetic fluids are outlined; the diffusion and magnetophoresis in dependence on the dimensions and magnetic moment of the particles of the disperse phase are indicated; the problems of stabilisation of the system and its aggregate stability is mentioned; the role of the size of the particles and the polydispersed nature of the composition in the formation of the magnetisation curve are outlined; the magnetisation curve is constructed and its application in magnetic granulometry discussed. Special attention is given to the features of the rheology of the fluid nano- and microdispersed magnetised media, using them as an example for introducing the concept of the 'Newtonian' and 'non-Newtonian' fluids. The effects specific for the nano- and microsize the disperse systems are discussed: magnetic levitation; magnetocaloric effect in the nanodispersedd magnetic system; 'suspension' of the magnetic fluid; kinetic and strength properties of the self-restored magnetic fluid membranes; the additive model of the elasticity of micro- and nanodispersedd systems taking into account interfacial heat exchange is presented, the magnetorheological effect is described. The properties of the magnetic dispersed system are determined to a large extent by the special features of the technology of production of these systems (the set of the initial components, the quality of these components, temperature conditions, the sequence of elements of the process, etc). Therefore, the textbook gives special attention to the physical-chemical aspect of the technology of producing magnetic fluids and ferrosuspensions. There are a large number of examples of calculating physical quantities which can used as a training material for obtaining the essential knowledge and self-control. The perception of the material is greatly facilitated as a result of using a large number of illustrations.

The authors, who prepared to this textbook, have considerable experience with the investigations in the area of the mechanics and acoustics of fluid nano- and microdispersed system, as indicated by the list of publication in specialised high-rating scientific journals in the last three years:

Polunin V.M., Storozhenko A.M., Ryapolov P.A., Tantsyura S.A. et al., The perturbation of the magnetization of the magnetic fluid by ultrasmall thermal vibrations that accompany the sound wave, *Akust. Zhurnal*, 2014, V. 60, No. 5, 476–483;

Polunin V.M., Ryapolov P.A., Karpova G.V., Prokhorov P.A., Oscillations of a bubble separated from an air cavity under compression caused by magnetic field in a magnetic Fluid, *Acoustical Physics*, 2014, V. 60, No. 1, 29–33;

Storozhenko A.M., Tantsyura A.O., Ryapolov P.A., Karpova G.V., Polunin V.M., Myo Min Tan, Interaction of physical fields under the acousto-magnetic effect in magnetic fluids, *Magnetohydrodynamics*, 2011. V. 47, No. 4, 345–358;

Polunin V.M., Storozhenko A.M., A study of the interaction of physical fields in the acousto-magnetic effect, *Akust. Zhurnal*, 2012, V. 58, No. 2. 215–221;

Polunin V.M., Boev M.L., Myo Min Than, Ryapolov P.A., Experimental study of an air cavity held by levitation forces, *Magnetohydrodynamics* 2012, V. 48, No. 3, 557–566;

Boev M.L., Polunin V.M., Lobova O.V., Fluctuations of the solitary bubble at the separation from the air cavity, compressed by the magnetic field in magnetic liquid, *Journal of Nano- and Electronic Physics*, 2013, V, No. 4, 04014 (5 pp);

Emelyanov S.G., Polunin V.M., Tantsyura A.O., Storozhenko A.M., Ryapolov P.A., From the dynamic demagnitizing factor to the heat capacity of a nanodispersed magnetic fluid, *Journal of Nano- and Electronic Physics*, 2013, V. 5, No. 4, 04027 (3 pp);

Boev M.L., Polunin V.M., Storozhenko A.M., Ryapolov P.A., Study of the interaction of physical fields in the acousto-magnetic effect for a magnetic fluid, *Russian Physics Journal*, 2012. V. 55,No. 5, 536–543;

Polunin V.M., Ryapolov P.A., Platonov V.B., and Kuz'ko A.E., Free oscillations of magnetic Fluid in strong magnetic field, *Acoustical Physics*, 2016, V. 62, No. 3, 313–318.

Three monographs were published:
Polunin V.M., *Acoustic effects in magnetic fluids*, Moscow, Fizmatlit, 2008;
Polunin V.M., *The acoustic properties of nanodispersed magnetic fluids*, Moscow, Fizmatlit, 2012;
Polunin V.M., *Acoustics of nanodispersed magnetic fluids*, New York - London, CRC Press, CISP Cambridge, 2015. - 472 p;

An article in the Great Russian Encyclopedia:
Polunin V.M., *Magnetic fluids*, Great Russian Encyclopedia, V. 18, Lomonosov – Manizer, Moscow, Great Russian Encyclopedia, 2011, 373–374.

Russian Federation patents:
Patent. RF 2507415, IPC F04B35/04, *Apparatus for compressing a gas by the working fluid*, **Shabanova I.A.**, et al.,
South Western State University, No. 2012123092/06; publ. 02/20/2014,
The method for determining the viscosity of the magnetic fluid or magnetic colloid, **Emelyanov S.G.**, et al., No. 2010112571/28; appl. 31.03.2010; publ. 10.04.2011. Bull. Number 10.

The generalization and presentation of the works performed to date on the mechanical properties of nano- and microdispersed magnetizing media and their application directly correspond to the programs and tasks of nanotechnological specialties and, therefore, the training manual of this orientation is necessary and timely.

Introduction to nanotechnology and microtechnology

1.1. Main concepts and definitions used in nano- and micro-technologies

The term '**nano**' originates from the Greek 'nanos' (dwarf) and corresponds to one billionth of a unity. The nanotechnologies and sciences of nanostructures and nanomaterials are concerned with objects with the size from 1 nm to 100 nm.

Nanotubes – fullerene tubes – tubulenes, represent hollow cylindrical formations assembled from hexagons with carbon atoms distributed at the vertices of the hexagons and usually having a spherical cap at the end, including pentagonal faces.

Fullerenes – molecules C_{60} and C_{70} having the form of a closed surface (spherical, ellipsoidal).

Magnetic fluids – disperse systems of different ferro- and ferrimagnetic single-domain particles in conventional fluids.

Suspension (suspensio in Latin) – a mixture of substances where the solid matter is distributed in the form of the finest particles in the liquid substance in the suspended (non-settled) state.

Ferrosuspensions (FS) – the disperse systems in which the disperse phase are the ferro- or ferrimagnetic particles.

The magnetorheological effect – a change of the rheological properties (viscosity, plasticity, elasticity) of some suspensions under the action of magnetic fields.

Magnetophoresis – directional magnetic diffusion under the effect of an inhomogeneous magnetic field.

Barodiffusion – gravitational diffusion which can be observed in the disperse system.

Sedimentation – settling of the particles of the disperse phase under the effect of the gravitational force.

The structure of the FS, induced by the external magnetic field, is responsible for the occurrence of specific rheological effects – viscoplasticity, viscoelasticity. Because of the complicated rheological specifications of such suspensions they are referred to as **magnetorheological suspensions**. The term 'magnetorheological fluid' relates also to fluids which solidify in the presence of the magnetic field. The particles in the magnetorheological fluid are basically of the micrometer range; they are very heavy for the Brownian motion to maintain them in the suspended state and, therefore, settle with time due to the natural difference in the density of the particles and the carrier fluid.

Cluster – a chemical compound containing a covalent bond between the atoms or molecules.

Lyophilic clusters can collect on the surface molecules of the surrounding medium and form strong solvate complexes with them.

Lyophobic clusters do not absorb the solvent molecules on their surface.

Colloidal clusters form in solutions as a result of chemical reactions and their size can be in the range from 1 to 100 nm.

The ligand (from the Latin word ligare – to bond) is an atom or molecule bonded with some centre – acceptor.

Mechanical dispersion – the process of mechanical refining of powder agglomerates in dispergators as a result of shear deformation

Adsorption – bonding of the substance (**adsorbate**) from a gas-like medium or from a solution with the surface layer of the fluid or solid (**adsorbent**).

Surface active substances (surfactants) – chemical compounds which concentrates on the interfacial surfaces of the thermodynamic phases and reduce the surface tension. The surfactant molecules usually contains the polar part (hydrophilic component).

Domains – the areas of spontaneous magnetisation of a ferromagnetic containing a large number of molecular magnetic dipoles, oriented parallel to each other. These formations reach the sizes of 10^{-5}–10^{-3} mm.

The magnetisation curve – the dependence of the magnetisation of substance M on the strength of the external magnetic field H.

The Curie point – temperature T_C at which the regions of spontaneous magnetisation (domains) breakup, and the ferromagnetics lose their magnetic properties.

The Langevin function $L(\xi)$ = cth $\xi- \xi^{-1}$ combined with the argument

$$\xi= \mu_0 m_* H/k_0 T$$

expresses the dependence of the magnetisation of the paramagnetic (superparamagnetic) M on the strength of the magnetic field H and temperature T:

$$M = nm_* L(\xi),$$

where m_* is the magnetic moment of the particle, n is the particle concentration.

The magnetic fluid seals (magnetic fluid gaskets) – the devices in which a magnetic fluid droplet overlaps the gap between a shaft and a sleeve as a result of the sustaining effect of the magnetic field, concentrated in the area of the gap.

Magnetic levitation – consequence of the effect of the ponderomotive force: the non-magnetic solids, situated in a magnetic fluid, placed in a magnetic field with a gradient along the direction of the force of gravity, are subjected to an buoyancy which can be many times greater than the weight of the displaced fluid.

The magnetocaloric effect – the variation of temperature of the magnetic substance during its adiabatic magnetisation or demagnetisation.

1.2. Position of the nano- and micro-objects on the scale of the sizes investigated at the present time

In order to compare efficiently the nanomaterials with the physical bodies and objects of the surrounding world on the basis of their size, it is possible to compare the Universe, the Earth and nanoparticles. The size of the observed Universe is estimated at 10–20 billion light years or $(1–2) \cdot 10^{26}$ m. The radius of the Earth is ~6370 km = 6.37 $\cdot 10^6$ m. Comparison shows how small are the dimensions of the objects of the nanoworld and nanotechnologies.

Each specific quantity is small or large only in relation to another quantity characteristic of the given conditions. The wavelength λ of light is small in comparison with the human height and, therefore, is ignored assuming that the light propagates in a straight line. In comparison with an individual atom λ is large and it is assumed that in interaction with the atom the amplitude of the light wavelength is identical at all points of the atom at any moment.

The temperature of 2000°C is too high in all areas of technology but in thermonuclear investigations the plasma with such a temperature is regarded as low-temperature plasma.

Table 1.1.

	Meter	Range
Macroworld	10^{25}	Size of the visible part of the Universe
	10^{24}	Distance between galaxies
	10^{21}	Size of the galaxies
	10^{18}	Interstellar distances
	10^{15}	Size of the solar system
	10^{12}	Size of the Earth
	$\sim 10^{5}$	Height of large mountains
	1	Height of man
	10^{-3}	Size of sand particle
	10^{-6}	Microscale $\sim 1-10$ μm Resolution of optical microscope
Nanoscale	10^{-7} to 10^{-9}	Submicrocrystalline materials ~ 100 μm Nanocrystalline materials ~ 10 nm Resolution of atomic force microscope ~ 5 nm Molecular clusters (fullerenes, etc), 3–4 nm 2–3 atomic molecules ~ 1 nm
One-particle world	10^{-10}	Size of the atom ~ 0.1 nm
	10^{-15}	Size of atomic nuclei
	10^{-19}	Studied structure of elementary particles

The character of the physical phenomena strongly depends on the size of the region of the space in which the phenomenon took place. Table 1.1 shows the scale of different quantities investigated by modern science.

One of the main characteristic quantities is the size of the atom, 10^{-10} m. Using this dimension, all the phenomena are divided into macroscopic and microscopic. The macrophenomena take place in the range $>10^{-7}$ m, and the microphenomena in the regions comparable with the atomic size of 10^{-10} m and smaller.

It should be noted that the concept 'the size of the atom' in the geometrical sense of the word has no meaning because the linear dimensions of the atom can be evaluated physically on the basis of the interaction between the atoms which is determined by the electromagnetic field of the atom with no sharp boundaries.

The directly observed bodies are macroscopic and consist of a large number of particles N. The large number of particles N is that for which the condition $\ln N \gg 1$ is fulfilled.

A very important parameter is the Avogadro number $N_A = 6.02 \cdot 10^{23}$ mole^{-1} which links the microscopic scale with the macroscopic one because the mole of any substance forms a body of the dimensions to which we are used to. For example, the mole of H_2O is $18 \cdot 10^{-6}$ m^3 of water.

The natural scale of velocity in nature is the velocity of propagation of light in vacuum $c = 2.998 \cdot 10^8$ m/s.

The Planck constant is also a universal constant and separates the laws of physics to quantum and classic $\hbar = 1.05 \cdot 10^{-34}$ J $\cdot$ s.

1.3. Prefixes to the units of the SI system

Any measurements (physical, technical, etc.) are based on physical laws, concepts and definitions. The physical, chemical and technical processes are determined by the quantitative data characterising the properties and state of the objects and bodies. To obtain these data, it is necessary to develop methods of measurement and the system of units. To measure any physical (technical and other) quantity, it means that this quantity must be compared with another homogeneous physical value regarded as the measurement unit (with a reference). Each physical quantity is a product of the numerical value by the measurement unit. Too large or small orders of numerical values (in relation to 10) are expressed in a shortened form by introducing new series of the units having the same name as the older ones but with a prefix.

To obtain this information in the area of synthesis of micro- and nanomaterials and investigate their properties, as the 'classic' physics, it is necessary to adhere to the systems of units and shortened terms developed in the world practice.

Table 1.2 shows the categories of the units and their shortened forms used in the scientific and technical literature throughout the world.

Table 1.3 gives the main and additional physical quantities and their measurement units in the SI system.

Table 1.2,

Prefix	Symbol Latin	Decadic logarithm	Prefix	Symbol Latin	Decadic logarithm
Tera	T	12	centi	c	−2
Giga	G	9	milli	m	−3
Mega	M	6	micro	μ	−6
Kilo	K	3	nano	n	−9
Hecto	H	2	pico	p	−12
Deca	Da	1	femto	f	−15
Deci	D	−1	atto	a	−18

Table 1.3.

Physical quantity	Symbol	Measurement unit	Units
Main measurement units			
Length	L	Meter	m
Mass	m	Kilogram	kg
Time	t	Second	s
Electric current intensity	I	Ampere	A
Temperature	T	Kelvin	K
Amount of substance	ν	Mole	mol
Light intensity	I_v	Candela	cd
Additional measurement units			
Flat angle	φ	Radian	rad
Solid angle	θ	Steradian	sterad

1.4. Influence of the dimensional effect on the physical properties of materials

At the start of the 20th century D. Thomson concluded that the observed anomalously high values of the electrical resistance of thin films in comparison with the course crystalline materials are linked with the restriction of the free path of the electrons by the size of that the specimen. The equation which proposed has the following form:

$$\frac{\rho_0}{\rho} = \frac{1}{2}k\left[\ln(k^{-1}) + \frac{3}{2}\right] \tag{1.1}$$

where ρ_0 is the specific electrical resistivity of a coarse-crystalline material, ρ is the specific electrical resistivity of the film; $k = \delta/l (k < 1)$, l is the free path of the electrons; δ is the film thickness.

The main special features of the manifestation of the dimensional effects in nanomaterials can be formulated as follows:
– decreasing grain size results in a rapid increase of the role of the interfaces;
– the properties of the interfaces in the nanometre range can differ from those in conventional coarse-crystalline materials;
– the size of the crystals can be compared, when their size decreases, with the characteristic dimensions of some physical phenomena (free path, the Broglie wavelength, etc).

The explanation of the physical nature of the dimensional effects is one of the most important problems of materials science. The examples of the effect of the dimensional factor on the physical properties of fluid disperse systems will be discussed later on.

1.5. The history of development of nanotechnologies and nano-objects

The beginning of investigations
The investigation of the small objects (powders, films, clusters, colloids) started long before the terms 'nanomaterials', 'nanotechnologies' were defined. Archaeological discoveries confirm the existence of procedures for preparing colloidal systems already in the classical antiquity. For example, the 'Chinese ink' was developed more than 4000 years ago, and the age of biological nanoobjects is attributed to the formation of life on the Earth.

It may be assumed that the first scientific mention of small particles was the disordered motion of the particles of flower dust on the surface of a fluid discovered in 1827 by the Scottish botanist R. Brown (R. Brown, Phil. Mag., 4, 161, 1828). The theory of

Brownian motion, developed in the first half of the 20th-century by A. Einstein, is used for determining the dimensions of the nanoparticles dispersed in fluids.

The start of the systematic studies of the nanostructured state of matter is believed to be the study of colloids and the formation of colloidal chemistry as an independent discipline in the second half of the 19th century. M. Faraday investigated in 1856–1857 the properties of colloidal solutions of highly dispersed gold and thin gold-based films. The variation of the colour in dependence on the particle size, observed by Faraday, is one of the first examples of investigation of the dimensional effects in the nano-objects.

Individual methods of the manufacturing and properties of ferro-suspensions were already known in the first half of the 20th century. The fluids with the suspended ferromagnetic particles from several micrometres to tens of micrometres in size had unique magnetic and rheological properties. The interest in the synthesis of media with their structure strongly affected by the magnetic field has been increasing rapidly with time.

The following main promising directions of the development of technologies can be specified: nanoprocessors with a low energy requirement and small dimensions; small memory devices with a very large memory capacity; the new medicinal preparations and methods of introducing them into the human body; new methods of environmental monitoring. The science of small objects is a combination of knowledge on the properties of substances and phenomena on the nanometre scale.

In 1990, IBM used a scanning tunnelling microscope to produce an abbreviation from 35 xenon atoms on the (110) face of a nickel single crystal. This demonstrated the possibilities of nanotechnologies.

The interest in this area of science is determined by three circumstances described below.

Firstly, the nanotechnologies can be used to produce completely new devices and materials, with the characteristics greatly exceeding the current level. This is very important for the rapid development of many areas of technology, biology, medicine, and environmental protection.

Secondly, the development of nanotechnologies takes place at the interface of different sciences and technologies of natural sciences (physics, chemistry, biology, computer technology, scanning microscopy, microelectronics, etc).

Thirdly, the introduction of disciplines dealing with nanomaterials and nanotechnologies into the education process has greatly expanded the base for the preparation of engineering personnel and experts with high qualification.

Many discoveries have been made in the 20th-century which have greatly changed the life of mankind. These discoveries include achievements in the area of nanotechnologies. Some of them will be briefly described.

The Josephson effect

In 1962, the English theoretical physicist Josephson (born in 1940) predicted the effect which opened new possibilities for high-accuracy physical measurements.

The principle of this effect is based on the phenomena accompanying the passage of superconducting current through a thin layer (~1 nm) of a dielectric material separating two semiconductors (Fig. 1.1). The thin layer can also be made from a semiconductor.

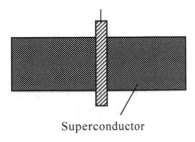

Superconductor

Fig. 1.1. The Josephson effect.

According to the superconductivity theory, the superconducting current is a flux of paired electrons. Electrons with opposite pulses and spins form pairs. When an external electrical field is applied the electron pairs move in the direction of the field with the velocity $\langle v \rangle = (v_1 + v_2)/2$. The distance between the electrons in the pairs is 10^{-6} m. Each pair, interacting with all other pairs, moves in a strictly matched manner. This results in the pulsed ordering in which the electron pairs have the same pulse. This means that all the pairs have the same wavelength $\lambda = h/p$.

The strict correlation of the pairs results in the situation in which the state of all pairs in the given superconductor is characterised by the same wave function $\Psi = |\Psi|e^{i\varphi}$, where φ is the phase of the wave function of the Cooper pair.

This function describes the propagation of the waves associated with movement of the particles. The wave function, like any other wave, has amplitude and phase. All the pairs are in the same state. They are matched in all parameters, including the phases.

The phase of the given superconductor is not determined. However, there is a completely determined phase difference in the range from 0 to 360° between the two superconductors. When the superconductors come closely together, their wave functions in the area of the gap start to overlap. The gap becomes a tunnelling contact. The exchange of the pairs between the superconductors starts to take place. The intensity and direction of the exchange are determined by the phase difference between these systems.

Prior to the Josephson discovery, it was assumed that there is no correlated tunnelling of the electron pairs. Josephson showed that because of the special properties of the superconducting state, the probability of tunnelling is the same as for the individual electrons.

The energy of two Cooper pairs on both sides of the transition differs by the value:

$$\Delta W = 2eU, \qquad (1.2)$$

here $2e$ is the charge of the pair, U is the potential difference.

This energy is received by the pair from the source in transition through the dielectric layer. This energy is excessive in relation to the energy of the ground state of the superconductor. Returning to the ground state, the electron pair radiates an energy quantum – phonon on with frequency v.

The portion of energy received by the pair $\Delta W = 2eU$ is emited in the form of a quantum

$$\hbar v = 2eU \qquad (1.3)$$

where $\hbar = 6.625 \cdot 10^{-34}$ J · s is the Planck constant.

From equation (1.3) one obtains

$$v = \frac{2eU}{h}. \qquad (1.4)$$

At $U = 1$ mV the frequency of alternating current is equal to 4.85 · 10^{11} Hz ($\lambda = 3.9$ mm).

There are two types of the Josephson effect – *stationary* and *non-stationary*. The stationary effect is found when the current through the junction does not exceed some critical value. In the stationary effect there is no voltage drop at the junction. The conduction electrons pass through the junction by the tunnelling effect.

If the current through the junction is higher than the critical value, the non-stationary effect is observed. In this case, a voltage drop U forms at the junction and the latter starts to emit electromagnetic waves with a frequency v.

The Josephson effect is used in superconducting interferometers containing two parallel Josephson junctions. The critical current for such a junction depends very strongly on the strength of the magnetic field so that the device can be used for very accurate measurements of the magnetic field.

Using the non-stationary Josephson effect it is possible to measure with high accuracy the electrical voltage and current intensity.

The Josephson contacts are used for reproducing with the highest accuracy the voltage reference 1 V.

The relative error is $5 \cdot 10^{-8}$.

When examining the Josephson effects, the fundamentals of nanotechnologies in the following directions were formulated:

1. Measurement of low currents (to 10^{-10} A);
2. Measurement of low voltages (up to 10^{-15} V);
3. Measurement of the induction of magnetic fields (up to 10^{-18} T);
4. Precision measurement of the constant $\hbar/e$: $\hbar/e = (4.135725 \pm 0.000026) \cdot 10^{-15}$ J·s/C.

Fullerite – the new form of carbon
The discovery of fullerenes – the new form of existence of one of the most widely encountered elements on the Earth, i.e. carbon – is regarded as one of the most surprising and important discoveries in the science of the 20th-century [8]. The history of discovery of the fullerenes started in 1973. Russian scientists D.A. Bochvar and S.G. Gal'pern published the results of quantum chemical calculations which indicated that there should be a stable form of carbon containing in the molecule 60 carbon atoms and having no substitutions. However, these proposals were greatly speculative, purely theoretical. It was difficult to confirm that such compounds

could be produced by chemical synthesis.

In 1985, R. Curl, H. Kroto and R. Smalley studied the mass spectra of graphite vapours produced under the impact of a laser beam and discovered that the spectra contain two signals whose intensity is considerably higher than that of all other signals. The signals corresponded to the masses 720 and 840 which indicated the existence of large aggregates of carbon atoms, C_{60} and C_{70}. The mass spectra can be used to determine only the molecular mass of the particle. The scientists proposed the structure of a polygon consisting of penta- and hexagons. This was the accurate repetition of the structure predicted earlier by D.A. Bochvar and S.G. Gal'pern.

Having the form of a closed surface, the C_{60} and C_{70} molecules were subsequently referred to as *fullerenes*. The modification of C_{60} carbon is the *fullerite*. The fullerenes can form compounds of different type and complexes with both simple elements and with their compounds. These materials were termed *fullerides*. Thus, a new area of physical materials science based on the new modification of carbon – fullerenes – was formed. The continued interest is maintained by prospects of using fullerenes in the nanoelectronics, power engineering and development of new polymers.

Fig. 1.2. C_{60} fullerene.

The fullerene C_{60} is a regular polyhedron with 60 carbon atoms in the vertices. It consists of 20 hexagons and 12 pentagons (Fig. 1.2). In contrast to the diamond, the fullerene is in fact a new form of carbon. The C_{60} molecule contains fragments with a fivefold symmetry (pentagons) which are banned by nature for inorganic compounds. Therefore, it should be accepted that the fullerene molecule is an organic molecule, and the crystal formed by such molecules (fullerite) is a molecular crystal which is the bonding link between the organic and inorganic substance.

The regular hexagons can be easily transformed to a flat surface, but they cannot be used to produce a closed surface. For this it is necessary to section part of the hexagonal rings and produce pentagons from the sections. In the fullerene the flat network of the hexagons (graphite network) is convoluted and crosslinked into a closed sphere. Part of the hexagons is transformed to pentagons. The resultant structure is referred to as a truncated icosahedron which has 10 axis of symmetry of the third order and six axis of symmetry of the fifth order. Each vertex of this figure has three closer's neighbours. Each hexagon borders with three hexagons and three pentagons, and each pentagon borders only with hexagons. Each carbon atom in the C_{60} molecule is situated in the vertices of two hexagons and one pentagon and is difficult to distinguish from other carbon atoms. The carbon atoms, forming the sphere, are bonded together with a strong covalent bond. The thickness of the spherical shell is 0.1 nm, the radius of the C_{60} molecule 0.357 nm. The length of the C–C bond in the pentagon is 0.143 nm, in the hexagon 0.139 nm.

The allied compounds and analogues of the fullerene are a few the moment. The best-known analogue C_{70} was produced almost simultaneously with C_{60}. The production of this analogue in the pure form is associated with considerable difficulties and, consequently, it has been studied far less frequently. Its form is close to the ellipsoid because of the slightly prolate shape (Fig. 1.3).

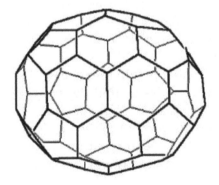

Fig. 1.3. C_{70} fullerene.

The C_{70} fullerene contains vertices of five types, for example, the vertices where three hexagonal faces converge. Two pentagonal faces are found on the prolate ends of the egg-shaped molecule. In their vicinity there are the bonds with the highest reaction capacity, with the properties similar to multiple bonds (two or three components).

Separation of a mixture of fullerenes, produced by evaporation of graphite, revealed the molecules C_{78}, C_{84} and also larger aggregates up to C_{200}. It is possible that other, larger analogues will also be produced. There are no theoretical restrictions for this.

A special group is formed by the so-called fullerene tubes – tubulenes, in the form of hollow cylindrical formations formed from hexagons and having in most cases a spherical cap at the end, including pentagonal faces (Fig. 1.4).

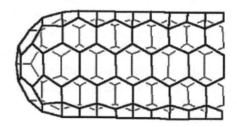

Fig. 1.4. Fullerene tubes – tubulenes.

These tubes form in the condensation of graphite vapours on a flat graphite substrate. The diameter of the tubes is 1–3 nm, the length reaches tens of nanometres.

In contrast to graphite and diamond, which have a crystal structure with a specific atomic order, the new structural element of carbon is the supermolecule. This means that the minimum element of the structure of the given modification of graphite is not atom but a multiatomic cluster.

The synthesis of carbon fullerenes (cage-like clusters of carbon C_n) and nanotubes – single- or multilayer cylinders, with the walls formed by the C_{60} hexagons – has opened the current stage in the investigation of the allotropy of carbon and initiated formulation of a large number of experimental and theoretical studies to find and produce new nanosized forms (nanoallotropes) of carbon.

Magnetic fluids

Magnetic fluids (MF) are colloids of various ferro- or ferrimagnetic single-domain particles in conventional liquid carrier media [1–6]. The magnetic fluids have a unique combination of the properties – high fluidity, capacity to magnetise to saturation, effective interaction with the magnetic field. The magnetic fluids were synthesised for the first time in the middle of the 60s of the 20th century, by producing nanoparticles of a solid magnetic material, dispersing the material in the carrier liquid and making the dispersion system aggregative stable – is one of the achievements of nanotechnologies. The magnetisation M of the concentrated magnetic fluids reaches ≈ 100 kA/m in the magnetic fields with a strength of $H \approx 80$ kA/m; in this case, their viscosity is close to the viscosity of the fluid–carrier and is almost completely independent of H.

The disperse medium is usually represented by magnetite Fe_3O_4, iron, cobalt, ferrites–spinels. The colloid of magnetite in kerosene, stabilised with oleic acid, was produced for the first time by Papell by dispersion in ball mills (S.S. Papell, US patent No. 3215572, 1965). The main type of magnetic fluid is the fluid based on magnetite, dispersed in hydrocarbon, organic silicon oil and in water. To prevent bonding (aggregation) under the effect of magnetic interaction, the particles are coated with one or two molecular layers of a surfactant (oleic acid, sodium oleate). At the average diameter of the magnetite particles of ≈ 10 nm, their magnetic moment is $\approx 2.5 \cdot 10^{-19}$ A $\cdot$ m², i.e. it is of the order of 10^4 atomic magnetic moments. Carrying out random thermal motion, the particles rotate through a large angle during the Brownian rotational period of approximately 1 μs at the viscosity of the fluid–carrier of 10^{-2} Pa $\cdot$ s.

Such small particles are sustained by the thermal Brownian motion in the volume of the fluid for as long as necessary. The high stability of the magnetic fluid is manifested in the magnetic fields with a strong inhomogeneity. The curve of the dependence of the static magnetisation of the magnetic fluid is similar to the Langevin function characterising the process of magnetisation of the paramagnetics. In the scientific literature, magnetic fluids are referred to as superparamagnetics.

The numerical value of the initial magnetic susceptibility χ of a concentrated magnetic fluid (volume concentration of magnetite of approximately 0.2) at room temperature reaches ~10, which is thousands of times greater than the susceptibility of the conventional fluids. When T approaches the Curie point T_C of the magnetic material from which the colloid is produced, its spontaneous magnetisation also shows a strong temperature dependence.

It should be noted that in the history of development of nanotechnologies there are also other suitable examples, including the development of materials characterised by the gigantic magnetoresistive effect, used widely in IT technology.

Questions for chapter 1

- *Please define the concept of the 'domain'. What does take place with the domains at the Curie point.*
- *What is the magnetorheological and magnetocaloric effect?*
- *What is magnetic levitation?*
- *What is the range of the linear dimensions of objects in nanotechnology? Which value can be regarded as the criterion for the ditinction between micro- and macro objects;*
- *What is the natural velocity reference?*
- *Which constant is used to distinguish between the quantum and classic physics?*
- *Specify prefices to the units of measurement of the physical quantities.*
- *When were the nanoparticles used for the first time? Which phenomenon can be regarded as the first scientific mention of small particles?*
- *Who synthesised the magnetic fluid for the first time?*

The physical model of the continuous medium

2.1. The continuity equation

The continuity equation expresses the law of conservation of matter in hydrodynamics: *the difference of the amount of the fluid, flowing in the given period of time into some volume and leaving this volume is equal to the increase of the amount of the fluid inside the given volume.*

We examine the elementary cubic volume

$$dV = dx \cdot dy \cdot dz \qquad (2.1)$$

shown in Fig. 2.1.

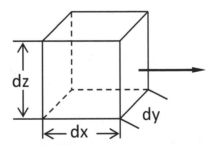

Fig. 2.1. The volume element.

Let it be that m_x is the mass of the substance transferred along the X axis during time dt.

At the point x:

$$m_{x_0} = \rho \cdot dy \cdot dz \cdot u_x \cdot dt \qquad (2.2)$$

At the point $x + dx$:

$$m_x = -(m_{x_0} + \frac{\partial m_x}{\partial x} \cdot dx) \qquad (2.3)$$

The increase of the mass in the investigated volume as a result of transfer along the X axis during the period dt:

$$\Delta m_x = -\frac{\partial m_x}{\partial x} \cdot dx = -\frac{\partial u_x \cdot \rho}{\partial x} \cdot dx \cdot dy \cdot dz \cdot dt \qquad (2.4)$$

Along the Y axis:

$$\Delta m_y = -\frac{\partial m_y}{\partial y} \cdot dy = -\frac{\partial u_y \cdot \rho}{\partial y} \cdot dx \cdot dy \cdot dz \cdot dt \qquad (2.5)$$

Along the Z axis:

$$\Delta m_z = -\frac{\partial m_z}{\partial z} \cdot dz = -\frac{\partial u_z \cdot \rho}{\partial z} \cdot dx \cdot dy \cdot dz \cdot dt \qquad (2.6)$$

the complete increment of the mass in the volume dV during the time dt is

$$\Delta m = -\frac{\partial u_x \cdot \rho}{\partial x} \cdot dx \cdot dy \cdot dz \cdot dt - \frac{\partial u_y \cdot \rho}{\partial y} \cdot dx \cdot dy \cdot dz \cdot dt -$$
$$-\frac{\partial u_z \cdot \rho}{\partial z} \cdot dx \cdot dy \cdot dz \cdot dt \qquad (2.7)$$

or

$$\Delta m = -\left(\frac{\partial (u_x \cdot \rho)}{\partial x} + \frac{\partial (u_y \cdot \rho)}{\partial y} + \frac{\partial (u_z \cdot \rho)}{\partial z} \right) \cdot dV \cdot dt$$

$$\Delta m = -\nabla (\vec{u} \cdot \rho) \cdot dV \cdot dt \qquad (2.8)$$

On the other hand, the decrease of the mass in the volume dV during the time dt can be described as

$$\Delta m = \frac{\partial \rho}{\partial t} \cdot dV \cdot dt \qquad (2.9)$$

Therefore, we obtain:

$$\frac{\partial \rho}{\partial t} \cdot dV \cdot dt = -\nabla(\vec{u} \cdot \rho) \cdot dV \cdot dt \qquad (2.10)$$

Consequently, the continuity equation for the fluid and the gas has the form:

$$\nabla(\rho\vec{u}) + \frac{\partial \rho}{\partial t} = 0. \qquad (2.11)$$

For a plane sound wave in a disperse system with the volume concentration of the particles of the disperse phase n the continuity equation has the form:

$$\frac{\delta n}{n} = -\frac{\partial u_d}{\partial x} \qquad (2.12)$$

where u_d is the displacement of the particles of the medium from the equilibrium position.

2.2. The equation of motion

In the mechanics of continuous media the fact that the medium consists of atoms and molecules is not taken into account. The volume examined in this case is small in comparison with the dimensions of macroscopic solids but sufficiently large in comparison with the distance between the molecules (in acoustics, the wavelength should be considerably greater than the average free path of the molecules).

We examine the equation of motion of the ideal medium (fluid) in which there is no viscous friction and no heat conductivity processes.

The motion of the fluid in a rectangular parallelepiped with the faces of the latter parallel to the coordinate axis will be examined (Fig. 2.2). The coordinates of one of the vortices (closest to the origin of the coordinates) will be denoted by x, y, z, the coordinates of the opposite vertex $(x + dx)$, $(y + dy)$, $(z + dz)$. The edges of the parallelepiped are consequently equal to dx, dy, dz.

The density of the fluid ρ depends on the coordinates x, y, z, and time t.

The force acting on the left face is:

$$F_1 = p \cdot dy \cdot dz \qquad (2.18)$$

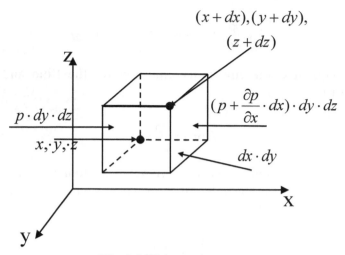

Fig. 2.2. Volume element.

The force acting on the right face is:

$$F_2 = \left(p + \frac{\partial p}{\partial x} \cdot dx \right) \cdot dy \cdot dz \qquad (2.14)$$

where p is the pressure in the fluid.

The algebraic sum of the projection of these forces on the axis x, y, z is equal to respectively

$$F_x = -\frac{\partial p}{\partial x} \cdot dx \cdot dy \cdot dz$$

$$F_y = -\frac{\partial p}{\partial y} \cdot dx \cdot dy \cdot dz \qquad (2.15)$$

$$F_z = -\frac{\partial p}{\partial z} \cdot dx \cdot dy \cdot dz$$

The investigated particles is subject to acceleration

$$\vec{a} = \frac{d\vec{u}(x, y, z, t)}{dt}.$$

In this case, it is necessary to use the total derivative of the complex function because the coordinates of the particle also depend on time, i.e.

$$\vec{a} = \frac{\partial \vec{u}}{\partial t} + \frac{\partial \vec{u}}{\partial x} \cdot \frac{\partial x}{\partial t} + \frac{\partial \vec{u}}{\partial y} \cdot \frac{\partial y}{\partial t} + \frac{\partial \vec{u}}{\partial z} \cdot \frac{\partial z}{\partial t}$$

We have

$$\vec{a} = \frac{\partial \vec{u}}{\partial t} + \frac{\partial \vec{u}}{\partial x} u_X + \frac{\partial \vec{u}}{\partial y} u_Y + \frac{\partial \vec{u}}{\partial z} u_Z \qquad (2.16)$$

On the basis of the second Newton law $\vec{F} = m\vec{a}$ or in the projections on the axis:

$$F_x = \rho \cdot dx \cdot dy \cdot dz \cdot a_x$$
$$F_y = \rho \cdot dx \cdot dy \cdot dz \cdot a_y$$
$$F_z = \rho \cdot dx \cdot dy \cdot dz \cdot a_z \qquad (2.17)$$

where ρ is the density of the medium.

Consequently, the projections of the acceleration vector on the coordinate axis are equal to

$$a_x = -\frac{1}{\rho} \frac{\partial p}{\partial x}$$

$$a_y = -\frac{1}{\rho} \frac{\partial p}{\partial y}$$

$$a_z = -\frac{1}{\rho} \frac{\partial p}{\partial z}. \qquad (2.18)$$

In the vector representation we have

$$\frac{\partial \vec{u}}{\partial t} + \frac{\partial \vec{u}}{\partial x} \cdot u_X + \frac{\partial \vec{u}}{\partial y} \cdot u_Y + \frac{\partial \vec{u}}{\partial z} \cdot u_Z = -\frac{1}{\rho} \left(\vec{i} \cdot \frac{\partial p}{\partial x} + \vec{j} \cdot \frac{\partial p}{\partial y} + \vec{k} \cdot \frac{\partial p}{\partial z} \right) \quad (2.19)$$

The resultant equation is the equation of motion of the ideal fluid.

The equation of motion for a viscous compressible fluid (Navier-Stokes equation) is the second Newton law for the elementary volume of the continuous medium:

$$\frac{\partial \vec{u}}{\partial t} + (\vec{u} \cdot \nabla)\vec{u} = -\frac{1}{\rho} \cdot \nabla p + \frac{\eta_s}{\rho} \cdot \Delta \vec{u} + \frac{1}{\rho} \cdot (\eta_v + \frac{\eta_s}{3})\vec{u} \cdot \nabla \vec{u} \qquad (2.20)$$

where η_S is the shear viscosity; η_v is the bulk viscosity.

It should be mentioned that the mathematical operators used in the equations (2.11) and (2.20) in the unfolded representation have the form

$$\nabla = \vec{i}\,\frac{\partial}{\partial x} + \vec{j}\,\frac{\partial}{\partial y} + \vec{k}\,\frac{\partial}{\partial z}$$

$$\Delta = \frac{\partial^2}{\partial x^2} + \frac{\partial^2}{\partial y^2} + \frac{\partial^2}{\partial z^2} \tag{2.21}$$

2.3. The equation of the mechanical state

The most important equations in the mechanics of the continuous media include the equation expressing the dependence of pressure in a substance on its density:

$$p = f(\rho, T). \tag{2.22}$$

This equation will be referred to as the equation of the mechanical state (in the theory of magnetism, as shown later, the equation of the magnetic state is used). The equation of the mechanical state for the ideal gases is the Mendeleev–Clayperon equation, and for the adiabatic process it is the adiabatic equation (Poisson equation).

The equation of the mechanical state for gases in the adiabatic process has the form

$$p \cdot V^\gamma = \text{const}, \tag{2.23}$$

were V is the volume of the gas, γ is the Poisson coefficient.

Since $\rho \sim 1/V$, we can write

$$p = \text{const} \cdot \rho^\gamma \tag{2.24}$$

In a sound wave, the non-perturbed value of the pressure p_0 and density ρ_0 increase slightly, δp and $\delta\rho$. From (2.24) it follows that

$$\delta p = \text{const} \cdot \gamma \cdot \rho^{\gamma-1}\, \delta\rho, \tag{2.25}$$

and consequently

$$\frac{\partial p}{p} = \gamma \cdot \frac{\partial \rho}{\rho}. \tag{2.26}$$

Thus, the relative increment of the pressure is proportional to the relative increment of density.

The universal equation of state cannot be obtained for the fluid systems. This is caused by the fact that the molecules are similar to each other and, consequently, the controlling role in the formation of the elastic properties of the fluid is played by the molecular interaction forces which have completely individual nature in every fluid. It is therefore necessary to use the empirical relationship between density ρ and pressure p. At small oscillation amplitudes the increment of the volume ΔV and pressure Δp (density $\Delta \rho$ or pressure Δp) are linked by the linear dependence

$$\frac{\Delta V}{V} = -\beta \cdot \Delta p; \qquad (2.27)$$

$$\frac{\Delta \rho}{\rho} = -\beta \cdot \Delta p, \qquad (2.28)$$

where the proportionality coefficient β is referred to as compressibility and has the dimension Pa^{-1}.

The relationships (2.27) and (2.28) can be regarded as the analogue of Hooke's law for fluids. The compressibility values differ for different fluids and are determined by experiments. Therefore, the equation of the mechanical state for the fluids written in the form

$$\delta p = \frac{1}{\rho \beta} \delta \rho, \qquad (2.29)$$

should be regarded as an empirical relationship. It should also be noted that for the fluids (with the exception of water) the value of β, obtained at a constant temperature β_T, is always higher than the value obtained in the absence of heat exchange β_S. In propagation of an ultrasound wave in a medium, the heat exchange between the adjacent compression and tension phases does not manage to take place. Therefore, the equation (2.29) uses in particular the adiabatic compressibility β_S:

$$\delta p = \frac{1}{\rho \beta_S} \delta \rho. \qquad (2.30)$$

In the elasticity theory the dependence $p = f(\rho)$ for solids is replaced by the equations linking the mechanical stress and tension (deformation) of the solid. The average longitudinal strain is the ratio of the elongation $\Delta \xi_{xx}$ to the initial length Δx:

$$\overline{S} = \frac{\Delta\xi_{xx}}{\Delta x}. \tag{2.31}$$

The longitudinal strain at the given point is the limit to which this relation tends at $\Delta x \rightarrow 0$, i.e.

$$S = \frac{\partial\xi_{xx}}{\partial x}. \tag{2.32}$$

The mechanical stress $\sigma(x)$ is the tension force per unit area of the cross-section of the rod.

The relationship between the stress and strain in the linear approximation is described by Hooke's law:

$$\sigma = ES, \tag{2.33}$$

where E is a constant characterising the material. It is referred to as the elasticity modulus (Young modulus).

The strain and stress propagate in the solid in the form of waves with the same velocity.

In a boundless isotropic solid, there can be longitudinal waves of the same type as in the fluid or gas, and also transverse (shear) waves. In solids of limited dimensions there may also be waves so other types (surface, bending). The possibility of propagation of waves of a specific type in a substance is directly linked with the existence of a specific type of elasticity (elasticity modulus).

If the lower face of the cube is 'fixed', then in order to deform it in the direction of the x axis (i.e., displace the upper face by $\Delta\xi_{xy}$), it is necessary to apply a tangential force (stress σ_{xy}) to the upper face of the cube (Fig. 2.3).

The Hooke law is fulfilled in the elastic deformation range

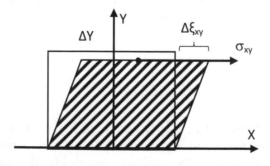

Fig. 2.3. Shear deformation.

$$\overline{\sigma}_{xy} = \mu \frac{\Delta \xi_{xy}}{\Delta y} \qquad (2.34)$$

or at $\Delta y \to 0$:

$$\sigma_{xy} = \mu \frac{\partial \xi_{xy}}{\partial y}. \qquad (2.35)$$

Coefficient μ is the shear modulus.

The moduli E and μ for the solids numerically differ from each other. Usually $E > \mu$.

In the bounded solids (in particular, in films) there may also be waves of other types propagating. Of these, one should mention in particular the surface waves (Rayleigh waves), normal waves in layers (Lamb waves), bending waves [7].

2.4. The elasticity coefficient

We will now discuss the oscillatory systems with concentrated parameters in which the active elements are nano- and microdispersed media [3]. One of the parameters of a similar oscillatory system is the elasticity coefficient which has different physical nature: the elasticity coefficient of the gas k_g, the elasticity coefficient of surface tension k_σ, the coefficient of ponderomotive elasticity κ_p. Coefficient κ_p will be examined in detail below.

The equation for calculating k_g will be derived on the basis of the equation of the state of gas (2.23). Coefficient k_g in the linear approximation is included in the equation for the Hooke law:

$$F = -k_g x, \qquad (2.36)$$

where x is the reduction of the length of the gas column in a pipe with diameter d after applying the pressure Δp.

On the other hand, force F can be described by the equation

$$F = \frac{\pi d^2}{4} \Delta p, \qquad (2.37)$$

where Δp is the excess pressure in the gas volume.

Taking into account the adiabatic nature of the process gives:

$$\Delta p = -\frac{\gamma P_0 \pi d^2}{4 V_0} x, \qquad (2.38)$$

where γ is the Poisson coefficient, P_0 is the gas pressure, V_0 is the volume of the gas cavity.

Consequently,

$$F = -\frac{\gamma \pi^2 d^4 P_0}{16 V_0} x. \tag{2.39}$$

Comparing the initial equation for the force with the resultant equation gives

$$k_g = \frac{\gamma \pi^2 d^4 P_0}{16 V_0}. \tag{2.40}$$

We take into account the well-known expression for the velocity of propagation of the sound wave in air

$$c = \sqrt{\frac{\gamma P_0}{\rho_g}}, \tag{2.41}$$

where ρ_g is the density of air, c is the speed of sound in air.

This gives

$$k_g = \frac{\rho_g c^2 S_0^2}{V_0}, \tag{2.42}$$

where S_0 is the circular cross-sectional area of the pipe.

The elasticity coefficient of surface tension k_σ is derived on the basis of the simplified calculation model shown in Figure 2.4.

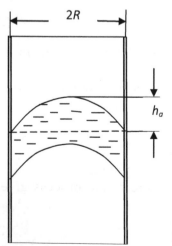

Fig. 2.4. Model for calculating k_σ.

The cross-section of a pipe with a radius R is overlapped by a plane-parallel layer of the magnetic fluid. It is assumed that taking into account the viscosity of the fluid on the side surface of the magnetic fluid membrane, the sticking condition is fulfilled and, consequently, the free surfaces of the membrane are periodically bent acquiring the shape of a spherical segment.

At small oscillations, the height of the sag h_a is small, i.e. $h_a \ll R$, and therefore the volume of the spherical segment can be presented in the form:

$$V_a = \frac{1}{6}\pi h_a \left(3R^2 + h_a^2\right) \approx \frac{1}{2}\pi h_a R^2. \qquad (2.43)$$

Taking into account the equality of the volumes

$$\pi R^2 \Delta Z = \frac{1}{2}\pi h_a R^2, \qquad (2.44)$$

where ΔZ is the displacement of the centre of gravity of the magnetic fluid membrane, we obtain:

$$h_a = 2\Delta Z. \qquad (2.45)$$

The work carried out by the increment of the free surface area of the magnetic fluid (both lower and upper) may be obtained from the equation

$$\Delta A = \sigma \cdot \Delta S = \sigma \cdot 2 \left[\pi\left(R^2 + h_a^2\right) - \pi R^2 \right] = 2\sigma\pi h_a^2 = 8\sigma\pi\left(\Delta Z\right)^2 \quad (2.46)$$

where σ is the surface tension coefficient of the magnetic fluid.

Considering the analogy with a spring pendulum gives the equation for the stiffness coefficient determined by surface tension:

$$k_\sigma = 16\,\pi\sigma. \qquad (2.47)$$

Thus, in the proposed model k_σ does not depend on the inner diameter of the pipe.

The coefficients k_g and k_σ are compared using the ratio k_σ/k_g, using the equations (2.42) and (2.47)

$$\frac{k_\sigma}{k_g} = \frac{64\sigma h_0}{\gamma d^2 p_0}. \qquad (2.48)$$

Let it be that $\gamma = 1.4$, $d = 1.5$ cm, $p_0 = 10^5$ Pa, $\sigma = 28 \cdot 10^{-3}$ N/m. To estimate k_g, the use a large volume of the gas cavity in order to obtain the minimum value of k_g with reference to the experiment conditions: $h_0 = 1$ m. Consequently, $k_\sigma / k_g \cong 0.05$.

Thus, the contribution brought by gas elasticity to the elasticity of the oscillatory system is almost two orders of magnitude greater than the contribution of surface tension.

2.5. The ponderomotive force

When a magnetic suspension moves in an inhomogeneous magnetic field, each particle with the magnetic moment $\vec{m}$ is subject to the effect of the force [1]

$$\vec{f} = \mu_0 (\vec{m}\nabla)\vec{H}. \tag{2.49}$$

The movement of a non-conducting magnetic fluid changes under the effect of the bulk magnetic force:

$$\vec{F} = \mu_0 (\vec{M}\nabla)\vec{H}, \tag{2.50}$$

which is derived from (2.49) by the summation: $\vec{M} = \sum\limits_{i=1}^{i=n} \vec{m}_i$ is the magnetisation of the system of the particles. Because at not too high frequencies $\vec{M} \uparrow\uparrow \vec{H}$, equation (2.50) can be replaced by

$$\vec{F} = \mu_0 M \nabla H. \tag{2.51}$$

The forces, acting on the magnetic from the side of the inhomogeneous magnetic field are referred to as the ponderomotive forces. In the case of non-conducting magnetised viscous liquid media it may be assumed that the electrical conductivity $\sigma \rightarrow 0$, i.e., the role of the bias currents is not large, then rot $\mathbf{H} = 0$. As a result of this, and also taking into account that $M - \rho\left(\dfrac{\partial M}{\partial \rho}\right)_{T,H} = 0$, the equation of motion has the following form:

$$\rho[\frac{\partial \vec{u}}{\partial t} + (\vec{u}\nabla)\vec{u}] = -\nabla p + \mu_0 (M\nabla)H + \eta_s \Delta \vec{u} + (\eta_\upsilon + \eta_s / 3)\nabla \operatorname{div}\vec{u}. \tag{2.52}$$

The equation of motion for an incompressible fluid has the following form:

$$\rho[\frac{\partial \vec{u}}{\partial t} + (\vec{u}\nabla)\vec{u}] = -\nabla p + \mu_0(\vec{M}\nabla)\vec{H} + \eta_s \Delta \vec{u}. \qquad (2.53)$$

In a number of cases, for example in acoustics, the second term in the square brackets may be ignored:

$$\rho\frac{\partial \vec{u}}{\partial t} = -\nabla p + \mu_0(M\nabla)H + \eta_s \Delta \vec{u} + (\eta_\upsilon + \eta_s/3)\nabla \mathrm{div}\,\vec{u}. \qquad (2.54)$$

The strength of the magnetic field is determined by the magnetostatics equations: $\mathrm{div}\mathbf{B} = 0$; $\mathrm{rot}\mathbf{H} = 0$.

In the presence of a system with an inhomogeneously magnetised fluid (a mixture of magnetic fluids of different concentration, the non-uniform distribution of air bubbles in the volume, non-uniform temperature distribution in the magnetic fluid) a magnetisation gradient ∇M forms in the magnetic fluid. The lowest value of the magnetisation of the magnetic fluid is found in the areas where its temperature is maximum, i.e., at the interface with the heat-exchanging surface. The ponderomotive force, determined by the magnetisation gradient ∇M, in a homogeneous magnetic field is determined by the equation:

$$F_M = \mu_0(\nabla M)M. \qquad (2.55)$$

The system of magnetohydrodynamics equations also includes the heat transfer equation which, according to the Ukrainian scientist I.E. Tarapov (1973), has the following form [9, 10]:

$$\rho T \frac{d}{dt}\left(S + \frac{\mu_0}{\rho}\int_0^H \left(\frac{\partial M}{\partial T}\right)_{\rho,H} dH\right) = \tau_{i,k}\frac{\partial v_i}{\partial x_k} + \mathrm{div}(\chi\nabla T) + (\mathrm{rot}\,\boldsymbol{H})^2/\sigma;$$

Here S is the entropy of the unit mass of the substance; χ is the heat conductivity coefficient; τ_{ik} is the viscous stress tensor; σ is the specific electrical conductivity of the medium. The remaining notations have the same meaning as previously. It is assumed that $\mathbf{B} = \mu_0 (\mathbf{H} + \mathbf{M})$ and $\mathbf{M}\|\mathbf{H}$, and the equation of state of the medium is written in the form:

$$p = f(\rho,T),$$

For the electrically non-conducting magnetic fluids it may be assumed that $\sigma = 0$. If the role of the bias currents is not large, then $\mathrm{rot}\mathbf{H} = 0$, and the heat transfer equation has the following form,

$$\rho T \frac{d}{dt}\left(S + \mu_0 \int_0^H \left(\frac{\partial M}{\partial T} \right)_{\rho,H} dH / \rho \right) = \tau_{i,k} \frac{\partial v_i}{\partial x_k} + \mathrm{div}\,(\chi \nabla T). \qquad (2.56)$$

Equation (2.56) differs from the appropriate equation for the 'conventional' non-magnetised liquid media by the second term in the round brackets of the left part which is a source of heat of magnetocaloric nature. The magnetocaloric effect under the normal conditions away from the Curie point for a dispersed magnetic is very weak and, therefore, this term in equation (2.56) can be ignored.

2.6. The magnetic pressure jump

A pressure jump forms at the interface of two media with different magnetic susceptibility when a magnetic field is applied [11]

$$\Delta p_H = -\frac{\mu_0 (\vec{M}_1 \vec{n})^2}{2} + \frac{\mu_0 (\vec{M}_2 \vec{n})^2}{2} \qquad (2.57)$$

where μ_0 with the magnetic constant, $\vec{n}$ is the unit vector of the normal to the interfacial surface, $\vec{M}_1$ and $\vec{M}_2$ are the magnetisation vectors of the media (index 1 relates to the 'lower' medium, index 2 to the 'upper' medium).

The pressure in the medium with a higher magnetisation is lower. If the upper medium is air, then the pressure inside the magnetic medium (in particular, the magnetic fluid) at the flat interface is lower in comparison with the atmospheric pressure by the value:

$$\Delta p_H = \frac{\mu_0 M_n^2}{2}, \qquad (2.58)$$

where M_n is the normal component to the magnetisation surface.

The physical nature of the pressure jump can be understood if it is taken into account that the jump of the normal component of the strength of the magnetic field forms at the interface of two media:

$$H_{2n} - H_{1n} = M_{1n} - M_{2n}. \qquad (2.59)$$

The magnetic micro- or nanoparticles, distributed on both sides from the interface, are situated in magnetic fields of different strength. In the vicinity of the interface there is a heterogeneous magnetic field whose gradient is directed from the medium with higher magnetisation to the medium with lower magnetisation

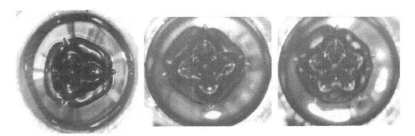

Fig. 2.5. Conical peaks on the surface of a magnetic fluid.

because in the latter the strength of the magnetic field is greater. The particle is subjected to the effect of the force $\vec{f} = \mu_0(\vec{m}\nabla)\vec{H}$ directed towards the less magnetic medium, in particular, into the atmosphere and this reduces the pressure in the magnetic fluid.

One of the manifestations of the magnetic pressure jump is the instability of the surface of the magnetic fluid in the magnetic field normal to it. This instability is based on the fact that when a certain critical strength of the magnetic field is reached (a certain level of magnetisation), the fluid surface assumes the specific needle-shaped form (a network of conical peaks). The formation of the peaks is due to the fact that a comparatively small perturbation of a smooth surface of the magnetic fluid distorts the magnetic field which causes their further growth.

A suitable example of the formation of conical peaks on the surface of the magnetic fluid is a photograph of the surface of the magnetic fluid membrane formed in a heterogeneous field of a ring-shape magnet [1]. A magnetic fluid droplet, introduced into the area with the maximum field in pipe, overlaps its cross-section. The magnetic field is normal to the free surface of the membrane.

Conical peaks form on both sides of the membrane. They are caused by the instability of the surface of the magnetic fluid in the transverse field. According to our observations, one to five approximately identical peaks can form, with the height of the peaks amounting to 1–2 mm. Figure 2.5 shows the photographs of the surface of a magnetic fluid membrane.

2.7. The mechanics of 'slipping' of nano- and microparticles in accelerated movement of the suspension

If a spherical particle is situated in a fluid carrying out translational movement, for example, oscillations with the oscillation velocity

$u = u_0 e^{-i\omega t}$, then at different densities of the particle and the liquid carrier the particle moves with the velocity W in relation to the fluid. This is caused by the property of inertia of the solids manifested under accelerated oscillatory movement. Consequently, denser solids (particles) lag behind particles with lower density (fluid) during movement.

The equation of the resistance force, acting on a sphere, has the form [12]:

$$F = 6\pi\eta R(1+\frac{R}{\delta})\dot{u} + 3\pi R^2 \sqrt{\frac{2\pi\rho}{\omega}}(1+\frac{2R}{9\delta})\frac{d\dot{u}}{dt}, \qquad (2.60)$$

here $\delta = \eta/\rho$ is kinematic viscosity.

The first term of the equation (2.60) is the Stokes drag assuming very slow oscillations. The second term is the dissipative force determined by the accelerated movement of the fluid flowing around the sphere. At $\omega = 0$ this equation changes to the Stokes equation. At high frequencies we obtain:

$$F = \frac{2\pi}{3}\rho R^3 \frac{d\dot{u}}{dt} + 3\pi R^2 \sqrt{2\eta\rho\omega}\,\dot{u}. \qquad (2.61)$$

The first term in this sum corresponds to the inertia forces in the laminar flow of the ideal fluid around the sphere, and the second term is the limiting expression for the dissipative force.

In the theory of disperse systems, developed by the Russian scientists S.M. Rytov, V.V. Vladimirskii and M.D. Galanin (1938), there is an equation for calculating the relative velocity of the particles in the medium $\tilde{\beta}_v$ ($\tilde{\beta}_v$ is the ratio of the oscillatory velocity of the suspended particles with the velocity of the surrounding medium) [13, 14]:

$$\tilde{\beta}_v = \frac{1+\sqrt{\Psi_v}+i\sqrt{\Psi_v}\left(1+2\sqrt{\Psi_v}/3\right)}{1+\sqrt{\Psi_v}+i\sqrt{\Psi_v}\left(1+b_2\sqrt{\Psi_v}\right)}, \qquad (2.62)$$

where $\Psi_V = \omega\rho_1 R_p^2 / 2\eta_{s1}$; $b_2 = \frac{2}{9}\left(1+2\frac{\rho_2}{\rho_1}\right)$; ρ_1 and η_{s1} are the density and shear viscosity of the liquid carrier; R_p is the particle radius of the disperse phase; ω is the circular frequency of harmonic oscillations; ρ_2 is the density of the particles of the disperse phase.

Separating the real part of the expression (4.3) leads to

$$\beta_v = \frac{\left(1+\sqrt{\psi_v}\right)^2 + \psi_v\left(1+2\sqrt{\psi_v}/3\right)\left(1+b_2\sqrt{\psi_v}\right)}{\left(1+\sqrt{\psi_v}\right)^2 + \psi_v\left(1+b_2\sqrt{\psi_v}\right)^2}. \tag{2.63}$$

Taking into account the possible slipping of the particles, the continuity equation has the form $\dfrac{\delta n}{n} = -\dfrac{\partial u_k}{\partial x}$, where u_k is the displacement of the particles from the equilibrium position. Since in a harmonic process $\vartheta = i\omega u$, $u_k = \dfrac{\tilde{\beta}_v \vartheta}{i\omega}$ then $\dfrac{\partial u_k}{\partial x} = \dfrac{\partial(\tilde{\beta}_v u)}{\partial x}$. The continuity equation has the following form in this case:

$$\frac{\delta n}{n} = -\frac{\partial(\tilde{\beta}_v u)}{\partial x}.$$

Figure 2.6 shows the graph of the dependence $\beta_v(R_p)$ on the semi-logarithmic scale (dependence $\beta_v(R_p)$ $[R_p]$ = m). It was assumed that $\rho_1 = 0.8\cdot10^3$ kg/m³, $\rho_2 = 5.2\cdot10^3$ kg/m³; $\eta_{s1} = 1.3\cdot10^{-3}$ kg/m·s; $v = 25$ MHz. The crosshatched area indicates the range of the values of R_p characteristic of the magnetic fluid (crosshatching on the left side). The graph shows that the slipping of the particles of the disperse phase is observed starting at $R_p \approx 450$ nm and when R_p increases to 1–10 µm it becomes large (the crosshatched area on the right). The range of the particle dimensions, corresponding to the stable magnetic fluid, is located at the start of the horizontal section of the curve.

Figure 2.6 shows that the 'slipping' of the nanoparticles in relation to the liquid carrier is almost negligible, whereas in the microsystems the intensity of this process will be very high.

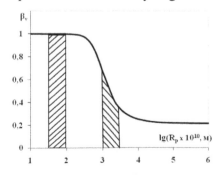

Fig. 2.6. $\beta_v(R_p)$ dependence.

Questions for chapter 2

- *Formulate the law of conservation of matter for the mechanics of continuous media.*
- *The equation of the movement of the ideal fluid, which assumptions are made when deriving the equation?*
- *Write the Navier–Stokes equation. Which process does this equation describe?*
- *What is the equation of mechanical state? What is its form for gases and fluids?*
- *What is the compressibility of the fluid and what are the units used for measurement;*
- *Which mechanisms of formation of elasticity operate in an oscillatory system with the active element in the form of a magnetic fluid;*
- *What is the ponderomotive force? Write the equation for calculating this force;*
- *Why the surface has the form of conical peaks in the magnetic field normal to the surface of the magnetic fluid;*
- *Is there any slipping of the particles in relation to the liquid matrix and the accelerated movement of the colloid; how does this process depend on the particle size?*

3

Measurement of magnetic parameters of nano- and microdispersed media

3.1. The magnetic field

We introduced the physical parameters representing the magnetic field and magnetisation of the material [15]. The magnetic field in vacuum is characterised by vector **H** – the strength of the magnetic field $[H]$ = A/m. The magnetic field in a substance is described by the magnetic induction vector **B**, $[B]$ = T (Tesla). In the case of weakly magnetic materials and also ferromagnetics with a low coercive force, these two parameters are linked together by a linear dependence

$$\mathbf{B} = \mu_0 \mu \mathbf{H} \qquad (3.1)$$

where μ_0 = 4 $\pi \cdot 10^{-7}$ A/m; μ is the magnetic permeability of the substance. For vacuum (and approximately also for air) μ = 1.

The magnetic moment of the unit volume of the substance is the magnetisation M, $[M]$ = A/m. Taking this parameter into account, we can write:

$$\mathbf{B} = \mu_0 (1 + \chi)\mathbf{H} = \mu_0(\mathbf{H} + \mathbf{M}), \qquad (3.2)$$

where χ is the magnetic susceptibility of the substance; $M = \chi H$ is the equation of the magnetic state.

The parameter χ changes its numerical value during magnetisation of the substance. In weak fields (at the beginning of the magnetisation curve) χ is the so-called initial susceptibility χ_0. The initial magnetic susceptibility χ_0 of specimens of a magnetic fluid can be determined from the initial slope of the magnetisation curve:

$$\mathbf{B} = \mu_0 (1 + \chi) \mathbf{H} = \mu_0(\mathbf{H} + \mathbf{M}), \tag{3.3}$$

When $\mathbf{H}$ increases, the ferromagnetic is magnetised to saturation, i.e., the value of M becomes maximum M_S for the given material (saturation magnetisation).

To characterise substances in magnetic fields of different strengths, the concept of the differential magnetic susceptibility is introduced: $\chi = \Delta M/\Delta H$. In the fields close to saturation $\chi \rightarrow 0$.

The magnetisation volume of strongly magnetic substances depends not only on the value of the magnetic susceptibility but also on their geometrical form. When magnetising a body which has finite dimensions and is made from a strongly magnetic material, in the external field, the magnetic poles ('magnetic charges' of the opposite sign) appear on both end surfaces of the body. This results in the appearance of a field with the opposite direction (demagnetising field H') in the material.

The strength of the field H' is proportional to the magnetisation M and, consequently, we can write

$$H' = N \cdot M, \tag{3.4}$$

where N is the proportionality coefficient which is called the demagnetising factor. The resultant field in the material H_i is described by

$$H_i = H_e - NM. \tag{3.5}$$

The exact and accurate calculation of the demagnetising factor is possible only for magnetics of the ellipsoidal shape [3].

In the transverse magnetisation of a long cylindrical bar the demagnetising factor is equal to:

$$N = 1/2. \tag{3.6}$$

In magnetisation of a bar along its axis, if the length is many times greater than the diameter, for example, a cylinder filled with a magnetic fluid, we have $N = 0$.

Assuming that $M(H) = \chi H_i$, we obtain $H_i = H_e - NM = H_e - N\chi H_i$ and, consequently

$$H_i = \frac{H_e}{1+N\chi} \quad \text{and} \quad M = \frac{\chi H_e}{1+N\chi}. \tag{3.7}$$

For a cylindrical column of a magnetic fluid in the magnetic field transverse to it $N = 0.5$ and therefore

$$H_i = \frac{H_e}{1+0.5\chi} \quad \text{and} \quad M = \frac{\chi H_e}{1+0.5\chi} \tag{3.8}$$

In the initial section of the magnetisation curve of the magnetic fluid $\chi = \chi_0 = $ const. In the non-linear region of the magnetisation curve it should be taken into account that $\chi = M/H_i$ – the ratio of the total (final) values of M and H_i. At $H_e \gg NM$ we obtain $H_i = H_e$ and, consequently, in a relatively strong magnetic field the demagnetising field can be ignored.

3.2. Description of experimental equipment and the method for getting the magnetisation curve

The strength of the magnetic field, strength gradient, and the magnetisation of the magnetic fluid and ferrosuspensions, and also the residual magnetisation of the FS are measured using different variants of the induction method. The induction method is based on the law of electromagnetic induction according to which the EMF of induction, formed in a conducting circuit, is numerically equal to the rate of change of the magnetic flux enclosed by the circuit [3].

To measure the magnetisation, the investigated fluid is poured into a cylindrical vessel. The longitudinal magnetising of the specimens is carried out either inside the solenoid calibrated with respect to current in advance, or between the pole terminals of the laboratory electromagnet. Magnetisation or remanent magnetisation is determined by the ballistic method by recording the change of the magnetic flux penetrating the turns of the measuring coil during withdrawal of the magnetised specimen from the coil.

In the simplest case, the magnetic flux penetrating the turns of the coil when switching on or switching off the external magnetic field is measured. Φ_M is the magnetic flux when the investigated specimen of the MF in inside and Φ_{MO} is the same without the MF. In this case, the magnetisation is calculated from the equation

$$M = (\Phi_M - \Phi_{MO})/\mu_0 \, SN_K \, (1 - N),\qquad(3.9)$$

where S is the area of the circuit, N_K is the number of turns in the coil, N is the demagnetising factor.

The changing of the magnetic flux in the coil for measuring the inductance (with and without the specimen) is also achieved by rotating the coil by 180° around the axis normal to the magnetic field lines.

The highest sensitivity is typical of the method of measurement based on using a system consisting of two inductance coils with anti-parallel connection. The vessel with the sample is placed inside one of the coils.

The diagram of experimental equipment for measuring the magnetisation of the magnetic fluid by this method is shown in Fig. 3.1. The vessel 1 filled with the sample of the magnetic fluid is placed inside one of the two identical inductance coils 2 and 3, connected in the opposite directions and fixed on the rotating rod 4. The measuring cell is then placed between the poles of the laboratory electromagnet 5 and connected to the microwebermeter 6.

When rotating the rod through 180° the magnetic flux changes as follows:

$$\Delta\Phi = \mu_0(M - (-M))S = 2\mu_0 MS,\qquad(3.10)$$

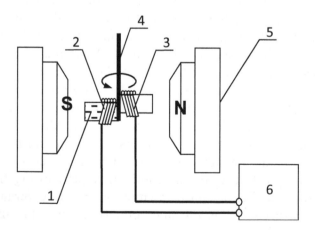

Fig. 3.1. Experimental setup for measuring magnetisation.

where $S = \pi d^2/4$, d is the inner diameter of the vessel.

The magnetisation is calculated from the equation:

$$M = 2\Delta\Phi\pi/\mu_0 d^2 \, N_{K1} \, (1-N) \qquad (3.11)$$

where N_{K1} is the number of turns in a single coil.

When the condition $d/l \ll 1$ is fulfilled, the demagnetising factor $N \approx 0$, and the relative measurement error ε_M

$$\varepsilon_M = \sqrt{\left(\delta\Delta\Phi/\Delta\Phi\right)^2 + \left(\Delta d/d\right)^2 + \left(\Delta N_{K1}/N_{K1}\right)^2} \qquad (3.12)$$

does not exceed 5%.

The saturation magnetisation of the magnetic fluid M_S can be determined from the limiting value of magnetisation in relatively strong magnetic fields required to reach this limiting value.

In accordance with the Langevin theory (section 8.1) the magnetisation of the magnetic fluid in the region of strong magnetic fields depends on the strength of the field H by the following equation:

$$M = M_S - \frac{3M_S k_0 T}{4\pi\mu_0 M_{S0} H R^3}, \qquad (3.13)$$

where M_{S0} is the saturation magnetisation of the dispersed ferromagnetic; R is the radius of ferroparticles.

The reliability and accuracy of measurement of M_S are higher if the resultant data are used to plot the graph of the dependence $M(H^{-1})$ and the linear approximation in the range $H^{-1} \approx 0$ is used.

3.3. The magnetisation curve

The magnetic parameters χ and M_S are determined by the ballistic method recording the magnetisation curve $M(H)$. The magnetic fluid fills a cylindrical vessel and the length of the vessel is considerably greater than its diameter so that the demagnetising field can be ignored. An example of analysis the magnetisation curve will now be described

Table 3.1 shows the physical parameters of the investigated sample of MF-01 which is a disperse system of magnetite in kerosene. The table uses the following notations: density ρ, solid phase concentration φ, initial magnetic susceptibility χ, saturation

Table 3.1

Sample	ρ, kg/m^3	φ, %	χ	M_S, kA/m	m_{*max} $\cdot 10^{19}$, A $\cdot$m^2	m_{*min} $\cdot 10^{19}$, A $\cdot$m^2	d_{max}, nm	d_{min}, nm
MF-01	1315	12	3.4	45.8	7.52	2.74	14.6	10.4

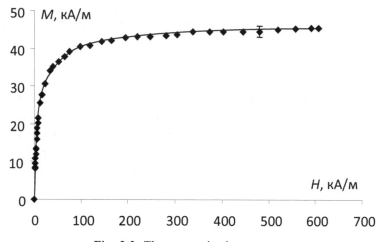

Fig. 3.2. The magnetisation curve.

magnetisation M_S. All the parameters were determined at a temperature of 31°C.

The resultant magnetisation curve of the magnetic fluid is shown in Fig. 3.2. In accordance with the Langevin function $L(\xi) = \text{cth}\,\xi - \dfrac{1}{\xi}$, $M = nmL(\xi)$, $\xi = \dfrac{\mu_0 m_* H}{k_0 T}$ the saturation magnetisation is reached at $\xi \sim 10$. Representing the magnetisation curve on a larger scale, it may be seen that its initial section is indeed straight. The initial magnetic susceptibility χ is determined by the slope of the initial section of the $M(H)$ curve.

To determine the saturation magnetisation M_S, i.e. the value of the magnetisation of the specimen at a very high ('saturating') strength of the magnetic field H, the following procedure is used in practice. In the vicinity of the values of $H^{-1} \approx 0$, the method of linear approximation of the $M(H^{-1})$ dependence is used to form a straight section and the required value of M_S is determined by extrapolating the section 2 to intersection with the coordinate axis (Fig. 3.3).

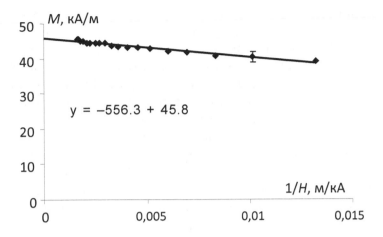

Fig. 3.3. Linear approximation of the $M(H^{-1})$ dependence.

3.4. Calculation of the 'maximum' and 'minimum' magnetic moments of nanoparticles and their diameters

To prevent the particles from deposition, the velocity of thermal motion of the particle should be greater than the settling velocity, determined by the Stokes equation. Consequently, the upper estimate of the diameter of the suspended particle is:

$$d \le \left(\frac{\eta^2 k_0 T}{\rho_s (\Delta\rho)^2 g^2} \right)^{1/7}, \qquad (3.14)$$

where $\Delta\rho = \rho_s - \rho_f$ is the difference of the densities of the solid and liquid phases.

Depending on the viscosity of the fluid η, equation (3.14) gives at room temperature $d_{max} \sim 10^{-5}$–10^{-6} m. In practice, in stable colloids the particle size is in the range 10^{-9}–10^{-6} m.

In the thermodynamically equilibrium condition, the distribution of the particles along the height is governed by the barometric law

$$n(z) = n_0 e^{-\frac{(\Delta\rho)Vgz}{k_0 T}}, \qquad (3.15)$$

where V is the volume of the particle.

When comparing the numerical value of the volume concentration φ, obtained from the density of the components of the system φ_ρ and from the saturation magnetisation $\varphi_M \equiv M_S/M_{S0}$, it appears that $\varphi_M/\varphi_\rho \approx 0.7$. One of the reasons for this difference is a small decrease of the size of the magnetic core of the particle as a result of the reaction of adsorption of surfactant molecules. For example, for the most widely used magnetic fluids with magnetite particles Fe_3O_4 the surfactant is oleic acid, and the mentioned effect results in the formation of the iron oleate which does not have any magnetic properties. Consequently, the 'magnetic diameter' of each particle decreases by the value of the double thickness of the non-magnetic layer $\varepsilon = 0.83$ nm (the constant of the crystal lattice of magnetite).

The concept of superparamagnetism was introduced in 1938 by W.C. Elmore. His experiments formed the basis of magnetic granulometry – the method of measuring the dimensions and magnetic moments of the fine particles using the magnetisation curve. A real disperse system is characterised by the distribution of the particles by the magnetic moments and linear dimensions. The estimates of the parameters of the nanoparticles, obtained by comparing the magnetisation curve with the ideal Langevin dependence, differ physically and quantitatively because they are produced by considering the initial and final sections of the magnetisation curve. The initial section of the magnetisation curve is formed basically by the particles with the largest magnetic moments, and the section of the vicinity of magnetic saturation is formed by the particles with relatively small magnetic moments.

Therefore, we can introduce conventionally the concepts of the 'maximum 'and 'minimum' magnetic moments of the nanoparticles [3].

The maximum and minimum magnetic moments of the nanoparticles m_{*max} and m_{*min}, obtained by the magnetic granulometry method, in accordance with the theory of superparamagnetism, are calculated from the equations

$$m_{*max} = \frac{3k_0 T \chi}{\mu_0 M_S}, \quad m_{*min} = \frac{k_0 T}{\mu_0 M_S \left(M/H^{-1} \right)}, \tag{3.16}$$

where M/H^{-1} is the tangent of the slope of the straight section of the $M(H^{-1})$ curve at $H \to \infty$.

The magnetic moment of the nanoparticles can also be described by the equation

$$\langle m_* \rangle = M_S/n,$$
$$\langle m_*^2 \rangle = 3kT\chi/\mu_0 n, \tag{3.17}$$

where n is the concentration of the particles.

The size of the particles (the diameter – assuming that the particles are spherical) is determined by the equation

$$d = \sqrt[3]{6m_* / \pi M_{S0}} = 0.016\sqrt[3]{m_*}, \tag{3.18}$$

where M_{S0} is the saturation magnetisation of the disperse phase (M_{S0} = 477.7 kA/m for magnetite).

The magnetic moment of the ferroparticles is expressed in terms of the saturation magnetisation of the ferromagnetic M_{S0}:

$$m_* = M_{S0} \cdot V_f,$$

where V_f is the volume of the magnetic part of the particle.

If the 'magnetic core' of a magnetite particles has the form of a sphere with a diameter d, then $m_* = \pi M_{S0}d^3/6 = 2.5 \cdot 10^5 d^3$. At d = 10 nm, $m_* = 2.5 \cdot 10^{-19}$ A·m².

Questions for chapter 3

• *Which physical parameters characterise the magnetic field? Which units are they measured in?*
• *Which physical parameters characterise the magnetic properties of materials? Name the measurement units.*
• *Why it is necessary to take into account the demagnetising field? How can this be done?*
• *Explain the nature of the ballistic method for producing the magnetisation curve of the magnetic fluid.*
• *Which equation describes the magnetisation of the magnetic fluid?*
• *How can the magnetisation curve be used to determine the initial magnetic susceptibility and saturation magnetisation?*
• *List the methods of evaluating the volume concentration of the disperse phase of a magnetic colloid? How do they differ?*
• *What is the magnetic granulometry? How can we determine the magnetic moment of the particles using the magnetisation curve?*
• *How are the magnetic moment and size of the particles connected? What is the range of dimensions of the particles of the magnetic fluids?*

4

The magnetocaloric effect in a nanodispersed magnetic system

The magnetocaloric effect is the change of the temperature of the magnetic material during its adiabatic magnetising or demagnetising [16, 17].

In an alternating homogeneous magnetic field the temperature of the magnetic fluid varies around the equilibrium value as a result of the magnetocaloric effect. Under the effect of thermal expansion the volume of the fluid also oscillates. The contribution of this process to the generation of elastic oscillations will be evaluated [3].

Let it be that the magnetic field whose time dependence is determined by the equation $H = H_o + H_m \cdot \cos \omega t$, is directed along the plane-parallel layer of the magnetic fluid. The amplitude of the alternating component of the magnetic field is so small that the condition $H_m \ll H_o$ is fulfilled. In the isothermal application of the magnetic field for which $H_n = 0$ there is no pressure gradient at the fluid–vacuum interface. However, in the adiabatic process a pressure jump forms at the magnetic fluid–vacuum interface and, as a result of this process, the fluid at the free interface is deformed. This will be demonstrated. The enthalpy differential E is represented in the form:

$$dE = T \cdot dS + V \cdot dp - \mu_0 V (\mathbf{M} \cdot d\mathbf{H}) \tag{4.1}$$

Consequently

$$\frac{(V-V_o)}{V} = -\mu_o \left\{ \frac{\partial}{\partial p} \left[V \int_0^H M \cdot dH \right]_{S,H} \right\} \Big/ V.$$

(4.2)

Here the index S denotes the adiabatic process in which there is no heat exchange, i.e., at constant entropy.

The amplitude of deformation of the fluid $\delta V_m/V$ when applying the field H_m can be determined as follows:

$$\frac{\delta V_m}{V} = -\frac{\mu_0}{V} \frac{\partial}{\partial p} \left\{ V \int_0^{H_0+H_m} M \cdot dH - \int_0^{H_0} M \cdot dH \right\}_S =$$

$$= -\frac{\mu_0}{V} \frac{\partial}{\partial p} \left\{ V \int_0^{H_0+H_m} M \cdot dH + \int_{H_0}^0 M \cdot dH \right\}_S =$$

$$= -\frac{\mu_0}{V} \frac{\partial}{\partial p} \left[V \int_{H_0}^{H_0+H_m} M \cdot dH \right]_S = \mu_0 \beta_S \left[\int_{H_0}^{H_0+H_m} M \cdot dH - \frac{\partial}{\partial p} \int_{H_0}^{H_0+H_m} M \cdot dH \right]_S$$

(4.3)

As a result of the inequality $H_m \ll H_0$ it can be assumed that $M = \mathrm{const} = M_o$ and, therefore,

$$\frac{\delta V_m}{V} = \mu_0 \left\{ \beta_S M_0 - \left[\frac{\partial M_0}{\partial p} \right]_S \right\} H_m.$$

(4.4)

We transfer from p to density ρ using the linear constitutive equation:

$$\frac{\delta V_m}{V} = \mu_0 \beta_S \left\{ M_0 - \rho \frac{\partial M_0}{\partial \rho} \right\}_S H_m.$$

(4.5)

In the absence of relaxation, the increment of magnetisation at the selected orientation of the vector $\mathbf{H}$ can be presented in the form:

$$\delta M_0 = -(n M_n + \gamma_* M_T) \frac{\partial u}{\partial x},$$

(4.6)

where n is the concentration of the magnetic particles (the number of particles in the unit volume), $M_n \equiv \left(\frac{\partial M}{\partial n} \right)_0$ and $M_T \equiv \left(\frac{\partial M}{\partial T} \right)_0$ relate to the unperturbed medium, $\gamma_* = qTc^2 C_p^{-1}$.

Using the continuity equation in the form of $\dfrac{\partial u}{\partial x} = -\dfrac{\delta \rho}{\rho}$, we obtain $\delta M = (nM_n + \gamma_* M_T)\dfrac{\delta \rho}{\rho}$ and consequently

$$\left(\frac{\partial M}{\partial \rho}\right)_S = (nM_n + \gamma_* M_T)/\rho. \tag{4.7}$$

Substituting expression (4.7) into (4.5) we obtain

$$\frac{\delta V_m}{V} = \mu_0 \beta_S \left\{ M_0 - nM_n - \gamma_* M_T \right\}_S H_m. \tag{4.8}$$

If $M_0 \sim n$, which is fulfilled, for example, for a system of ferroparticles with the magnetisation of the system described by the Langevin equation, then $nM_n = M_0$ and

$$\frac{\delta V_m}{V} = -\mu_0 \beta_S \gamma_* M_T H_m. \tag{4.9}$$

The expression for the magnetocaloric effect, well-known in the theory of magnetism, is:

$$dT = -\mu_0 T(M_T)_H \frac{dH}{\rho C_p}. \tag{4.10}$$

Therefore, the equation (4.8) can be presented in the following form

$$\frac{\delta V_m}{V} = -\mu_0 T M_T q H_m / \rho C_p = q \cdot dT. \tag{4.11}$$

In the alternating magnetic field, the magnetocaloric effect and the thermal expansion typical of the fluid result in the oscillations of the volume of the fluid with the frequency of the field oscillations.

The magnetocaloric effect fulfils in this case the function of one of the possible mechanisms of electromagnetic excitation of elastic oscillations in the magnetic fluid. The dependence of the oscillation amplitude on H is determined by the multiplier M_T in equation (4.11). Assuming that the magnetising of the fluid takes place in accordance with the Langevin function, we obtain

$$M_T = (M_S / T)(\xi sh^{-2}\xi - \xi^{-1}) \tag{4.12}$$

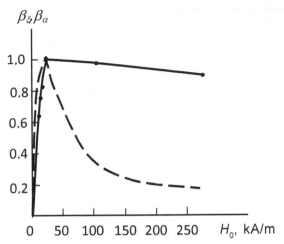

Fig. 4.1. Dependence of the amplitude of excited oscillations on the strength of the magnetic field.

and, consequently

$$\frac{\delta V_m}{V} = \mu_0 q H_m M_S D(\xi) / \rho C_p \qquad (4.13)$$

where $D(\xi) = \xi^{-1} - \xi \, \text{sh}^2 \, \xi$.

Figure 4.1 shows in the relative units the field dependences of the amplitude of elastic oscillations of the magnetic fluid: i) β_ξ (dashed line) obtained on the basis of the Langevin theory (section 8.1) and ii) β_a (solid line) using the experimental data β_a (solid line).

The $\beta_\xi(H)$ dependence was constructed assuming that $m_* = 2.5 \cdot 10^{-19}$ A · m^2 and $T = 290$ K. Function $D(\xi)$ has a single maximum at the point $H = 25.41$ kA/m, and its numerical value is 0.348.

The amplitude of deformation of the fluid $\delta V_m/V$ can be estimated using the values usual for the standard kerosene magnetic fluid: $D(\xi) = 0.35$, $M_S = 50$ kA/m, $H_m = 1.5$ kA/m, $q = 0.64 \cdot 10^{-3}$ K^{-1}, $\rho = 1250$ kg/m^3, $C_p = 2100$ J/kg·K which will be substituted into equation (4.13), consequently, we obtain $\delta V_m/V = 0.8 \cdot 10^{-8}$.

However, the experimental value of static deformation $\Delta h/h$ is $0.8 \cdot 10^{-7}$ which is an order of magnitude greater than the estimate of deformation obtained as a result of the magnetocaloric effect. The solid line in Fig. 4.1 shows the experimental curve of the dependence of the relative amplitude β_a on the strength of the magnetic field H obtained using the data for a specimen of the magnetic fluid based on magnetite and MVP oil at a frequency of 2 MHz. The positions of the maxima of the theoretical and experimental curves are close to each other but their falling branches greatly differ.

Evidently, the magnetocaloric effect in the magnetic fluid based on magnetite cannot play the key role in the electromagnetic excitation of elastic oscillations in the megahertz frequency range.

In addition to the release or absorption of heat in remagnetisation of the fluid, which is determined by the alignment of the ferromagnetic dipoles in the fields, the release or absorption of heat caused by the *intrinsic* magnetocaloric effect of the ferromagnetic phase also takes place. The contribution of this effect to the change of the volume of the fluid will be evaluated.

The amount of heat, generated during the magnetising of a single particle Q_{T1} can be determined using the expression (4.10) which after transformations has the form

$$Q_{T1} = -\mu_0 V_G T (\partial M_G / \partial T)_H \cdot H_m \qquad (4.14)$$

where G is the index of the solid ferromagnetic.

The volume of the solid phase of the magnetic fluid is $V_G = \varphi V$ and therefore the heat, released in the volume of the dispersed medium, can be determined from the expression

$$Q_{T1M} = -\mu_0 \varphi V T (\partial M_G / \partial T)_H \cdot H_m. \qquad (4.15)$$

Therefore, in calculation per unit mass of the fluid

$$Q_{T1M} = -\mu_0 \varphi T (\partial M_G / \partial T)_H \cdot H_m / \rho. \qquad (4.16)$$

The expression (4.16) can be used to determine the increase of the temperature of the fluid:

$$\partial T_M = Q_{T1M} / C_{pH} = -\mu_0 \varphi T (\partial M_G / \partial T)_H \cdot H_m / \rho C_{pH}, \qquad (4.17)$$

were C_{pH} is the specific heat at $p = $ const and $H = $ const.

The relative increase of the volume is:

$$\delta V_m / V = -\mu_0 \varphi q T (\partial M_G / \partial T)_H \cdot H_m / \rho C_{pH} =$$
$$= \rho_G C_{pG} \varphi q \cdot \delta T_G / \rho C_{pH}. \qquad (4.18)$$

In the ferromagnetics, the maximum magnetocaloric effect is recorded at the Curie point T_C. For example, if the magnetic phase is represented by gadolinium Gd, for which $T_C = 293$ K, then we can obtain a large magnetocaloric effect in the vicinity of room temperature. According to the data obtained by K.P. Belov et al, at $H_0 = 200$ kA/m, $\Delta T/\Delta H = 0.25 \cdot 10^{-5}$ K $\cdot$ m/A, $\rho = 7.98 \cdot 10^3$ kg/m^3, $C_{pG} = 320$ J/(kg $\cdot$ K). Consequently, at $H_m = 1.5$ kA/m $\Delta T_G = 3.75 \cdot 10^{-3}$ K and $\partial V_m/V = 0.35 \cdot 10^{-6}$, which is almost 2 orders of magnitude greater than the result obtained for the magnetocaloric effect of alignment of the dipoles in the fluid based on magnetite and kerosene.

Questions for chapter 4

- *What is the magnetocaloric effect?*
- *Estimate the amplitude of deformation of the magnetic fluid when an external magnetic field is applied.*
- *Describe the contribution to the change of the volume of the fluid by the intrinsic magnetocaloric effect of the ferromagnetic phase.*
- *Can the magnetocaloric effect play the key role in the electromagnetic excitation of elastic oscillations in the magnetic fluid? Why?*

5

Effect of the ponderomotive force

5.1. Experimental confirmation of the ponderomotive mechanism of electromagnetic excitation of elastic oscillations in a magnetic fluid

The ponderomotive mechanism of electromagnetic excitation of elastic oscillations in the magnetic fluid was initially regarded as the only possible mechanism. Formally, this conclusion was made from the main equation of the quasi-equilibrium hydrodynamics of magnetic fluids where the irreversible process of magnetising can be ignored and $\mathbf{M} \uparrow \uparrow \mathbf{H}$:

$$\rho \frac{d\vec{v}}{dt} = -\nabla p + \eta \Delta \vec{v} + \mu_0 M \nabla H, \qquad (5.1)$$

where the strength of the magnetic field is determined by the magnetostatics equations: $\mathrm{div}\mathbf{B} = 0$; $\mathrm{rot}\mathbf{H} = 0$.

There is only a small number of studies concerned with the examination of the physical nature of the mechanism of generation of elastic oscillations by the magnetic fluid located in an alternating magnetic field with the oscillation frequency of several tens of kilohertz. Therefore, it became necessary to investigate by experiments this question (V.M. Polunin, N.M. Ignatenko, V.A. Zraichenko, 1990 [18]).

Figure 5.1 shows the diagram of the experimental setup for the examination of the magnetoacoustic effect in a magnetic fluid in the range of low ultrasound frequencies.

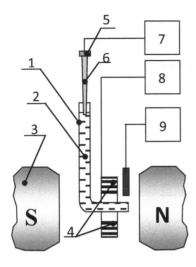

Fig. 5.1. Diagram of experimental setup.

The glass L-shaped tube 1 is filled with the magnetic fluid 2 to be investigated. The lower horizontal nozzle is situated between the poles of the laboratory electromagnet 3 producing a constant homogeneous field. The exciting inductance coil is inserted on the nozzle in the coaxial direction. An air gap approximately 2 mm wide is formed between the coil and the glass tube. The presence of a standing elastic wave in the *magnetic fluid–cylindrical tube* system is fixed by the piezoelectric sheet 5 situated on the end of the metallic bar-waveguide 6. The oscilloscope 7 is designed for determining the alternating electrical voltage taken from the piezoelement; 8 is the alternating voltage generator; 9 is the magnetic induction measurement device.

The experiments were carried out using a magnetic fluid based on kerosene with the density of $\rho = 1300$ kg/m^3 and the saturation magnetisation of $M_S = 51$ kA/m.

Figure 5.2 shows the dependence of the relative amplitude β_e of the exciting oscillations on the strength of the direct component of the magnetic field H_0, produced at a frequency of 20 kHz. The crosshatched circles are the values obtained when increasing the strength of the magnetic field, the open circles – when decreasing the strength. The same figure shows the results of measurements of $\beta_M = M/M_{max}$ (crosshatched rhombs, the dotted curve) at $M_{max} = 49$ kA/m. In the range of technical saturation both curves are qualitatively identical – they reach saturation with increasing strength of the field.

The ponderomotive mechanism, as indicated by the relationship $f_p = \mu_0 MV\nabla H$, determines the effect of the driving force, proportional

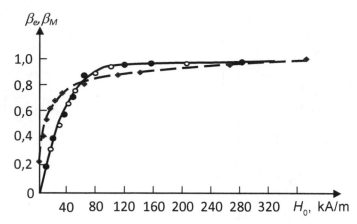

Fig. 5.2. Dependences: $\beta_e(H_0)$ – solid curve, and $\beta_M(H_0)$ – dotted curve.

to the magnetisation of the fluid, as also observed in the experiments described here.

The ponderomotive mechanism is also characterised by the linear dependence of the amplitude of the generated sound on the amplitude of the alternating magnetic field.

In the investigated frequency range, the presence of such a dependence is confirmed in the experiments carried out in the experimental setup shown in Fig. 5.3. The exciting inductance coil (inductor) 2 is connected with the oscillator 1. The coil is placed inside the ring-shaped magnet 3. The movement of the magnet is recorded with the cathetometer 4. A standing wave forms in the

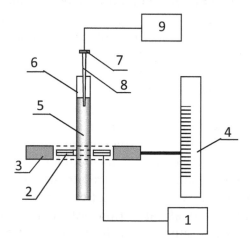

Fig. 5.3. Diagram of equipment.

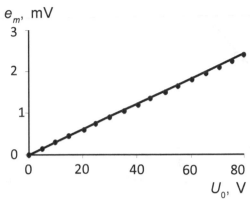

Fig. 5.4. The $\beta(U_0)$ dependence.

magnetic fluid 5 filling the cylindrical shell 6. The presence of the wave is recorded using the piezoelectric sheet 7 placed at the end of the waveguide. The oscilloscope 9 is designed for observing and measuring the electrical signal.

If the inductance coil (inductor) is the source of the alternating magnetic field, the amplitude of the strength of the magnetic field H_m, according to the Biot–Savart–Laplace law, is directly proportional to the amplitude of current intensity I in the inductor. The same dependence on the current intensity I is observed for ∇H_m and the ponderomotive force f_p.

Figure 5.4 shows the results of measurements of the amplitude of the oscillograms e_m for different values of the amplitude of the alternating voltage U_0 and the approximating curve.

The linear dependence of the relative amplitude of the oscillations of magnetisation of the magnetic fluid β (M) is confirmed by the graph shown in Fig. 5.5. The experimental data were obtained at a frequency of 24.2 kHz.

Thus, the assumption of the dominating role of the ponderomotive mechanism of electromagnetic excitation of the sound oscillations in the magnetic fluid, used often in different theoretical models, does not contract the experimental data obtained in the kilohertz frequency range. In the following examination of specific cases of the electromagnetic excitation of elastic oscillations in the magnetic fluid attention will be given to the concept of the ponderomotive mechanism.

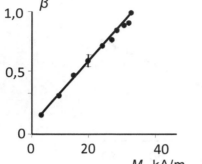

M, kA/m   **Fig. 5.5.** The β(*M*) dependence.

5.2. The ponderomotive mechanism of excitation of oscillations in a cylindrical resonator with a magnetic fluid

The effect of the electromagnetic field on the magnetic fluid may result in the formation in the fluid of oscillations of different type: elastic, surface, oscillations of the shape. A special position amongst them is occupied by the elastic oscillations – sound and ultrasound oscillations which are of obvious scientific interest. The transformation effect of this type will be referred to as the magnetoacoustic effect (MAE).

In the applied aspect, the magnetic fluid in this problem plays the role of a material used for transforming the energy of the electromagnetic field to the energy of elastic oscillations. The transforming devices have a number of advantages in comparison with the traditional solid-state magnetostriction and piezoelectric transducers. These advantages are based on the following: in comparison with the solid state transducer the working body of the transducer has a lower density and the speed of sound and its mass is more than an order of magnitude smaller than the mass of the emitter for the same resonance frequency; the surface of the magnetic fluid is capable of acquiring any geometry, specified by the shape of the container; the resonance frequency and the directional diagram can be smoothly regulated; the quality of the wave resistances of the magnetic fluid and seawater predetermine the possibility of constructing a wide-band source of elastic oscillations.

The first attempts to solve this problem were based on the use of 'coarse' ferrosuspensions. However, because of the rapid delamination of these systems and the considerable damping of elastic oscillations in them, the transducers of this type could not be used. These shortcomings are not found in the transducers in which the magnetic fluid is the active element. The problems of the electromagnetic excitation of the acoustic oscillations in the magnetic fluids were examined for the first time theoretically by Cary and Fenlon. The active element of the transducer examined by these authors had the shape of a cylindrical disc – tablet. The case of an infinite plane-parallel layer was investigated; the external magnetic field was directed along the normal to the layer. The results of thermodynamic transformations show that a pressure gradient, determined by the 'jump' of the strength of the magnetic field, forms at the boundaries of the layer. The functioning of only one ponderomotive mechanism is permitted in the region of technical saturation, and this should help in producing a source of oscillations competing with the conventional magnetostriction and piezoelectric transducers in the range 100–150 kHz in which the non-conducting magnetic fluids are characterised by small losses due to eddy currents and remagnetisation.

The first results of experimental studies of the special features of functioning of the magnetic fluid transducers were obtained by the Belarusian scientists A.R. Baev, G.E. Konovalov and P.P. Prokhorenko at frequencies of 16–27 kHz [19].

When the magnetic fluid fills the cavity with a specific geometry – the resonator – it is possible to use the simplest source of the magnetic field with its main role being the formation of the alternating component of the field. The resonance excitation of the oscillations is obtained by selecting the frequency of variation of the driving force.

This problem was solved in studies by V.M. Polunin (1978) in which attention was given to a cylindrical model of the magnetic fluid emitter (MFE) [20, 21]. In the simplest case this model may be produced by immersing a straight conductor through which alternating current flows, in the magnetic fluid.

In the theoretical aspect, the advantage of this model is that its analysis does not require the use of any empirical or semi-empirical equation determining the geometry and time dependence of the magnetic field. The examined magnetic field is the field of a straight infinite conductor with the current, and its geometry is well-known and it is determined by the Biot–Savart–Laplace law. The principle of the method is investigated on the model shown in Fig. 5.6.

This method of excitation of the resonance oscillations in the magnetic fluid is based on the application of a heterogeneous magnetic field containing a component varying with time according

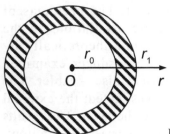

Fig. 5.6. Cylindrical resonator.

to the harmonic law. The magnetic fluid fills the space between the coaxial infinitely long cylindrical surfaces with the radii r_0 and r_1. The proposed model uses a non-viscous, homogeneous fluid with no thermal conductivity. The special feature of magnetising of this fluid will be mentioned below.

The cylinders, restricting the fluid, are regarded as non-magnetic, with no electrical conductivity and absolutely rigid. The assumption of the electrical and magnetic 'neutrality' of the restricting cylinders is the reason why the electromagnetic induction in them and all consequences of it can be ignored.

In the cylindrical coordinate system using the experiments, the Z axis coincides with the axis of the cylinders and is directed to the drawing. The wave equation, describing the displacement of the fluid particles from the equilibrium position in the cylindrical coordinates has the form

$$\frac{\partial^2 u}{\partial t^2} = c^2 \left(\frac{1}{r} \cdot \frac{\partial}{\partial r} r \frac{\partial u}{\partial r} \right) + F_H, \tag{5.2}$$

where $u(r, t)$ is the displacement of the particles from the equilibrium position; r is the coordinate; t is time; c is the speed of sound in the fluid which is an actual value because there are no dissipative processes determined by the viscosity or thermal conductivity of the medium; F_H is the driving force which in the investigated model is linked with the heterogeneity of the magnetic field, generated by the axial current in the medium with no electrical conductivity.

The right-hand part of equation (5.2) is the ratio of the elasticity force, acting on the elementary volume of the fluid dV, to the mass of the fluid in this volume. When applying the heterogeneous magnetic field to the volume dV of the magnetic fluid with no electrical conductivity we obtain the ponderomotive force $d\mathbf{F}$ whose magnitude and direction can be determined on the basis of the quasistatic hydrodynamics of the isotropic magnetic fluid using the equation:

$$\mathbf{F} = \mu_0 M \nabla H/\rho. \qquad (5.3)$$

The driving force in the investigated model depends strongly on the type of function $\mathbf{M(H)}$, i.e., on the specific type of the equation of the magnetic state which in turn should satisfy the Maxwell equations $\text{div}(\mathbf{H+M}) = 0$ and $\text{rot}\mathbf{H} = 0$.

This requirement is satisfied in particular by the linear equation

$$M = \chi H, \qquad (5.4)$$

where χ is magnetic susceptibility, and the magnetic saturation equation

$$\mathbf{M} = \mathbf{M}_s = \text{const.} \qquad (5.5)$$

In the first case, the mass unit will be under the effect of the additional force $\mathbf{F}_M$ equal to

$$\mathbf{F}'_M = \mu_0 \chi \nabla \mathbf{H}^2/2\rho. \qquad (5.6)$$

In the second case

$$\mathbf{F}''_M = \mu_0 M_S \nabla H/\rho. \qquad (5.7)$$

To ensure the cylindrical symmetry of the solved problems, it is necessary to use a magnetic field generated by an infinite conductor with current directed along the *OZ* axis. The time dependence of current *I* is given in the form

$$I = I_0 + I_m \cos \omega t, \qquad (5.8)$$

where I_0 and I_m are the constant component of current and the amplitude of the alternating component of the current, respectively; ω is the circular frequency of alternating current.

On the basis of the quasistatic approximation using the Biot–Savart–Laplace law, the following equation is obtained for **H**:

$$\mathbf{H} = (2\pi r)^{-1}(I_0 + I_m \cdot \cos \omega t)e_\varphi \qquad (5.9)$$

where e_φ is the unit vector.

Substituting (5.9) into (5.5) and (5.7), we obtain the projection of the force $\mathbf{F}'_M$ and $\mathbf{F}''_M$ on the direction of the unit vector:

$$F'_M = -\mu_o \chi I_m^2 (1 + \cos 2\omega t) / 8\pi^2 r^3 \rho; \qquad (5.10)$$

$$F''_M = -\mu_o M_s (I_0 + I_m \cos \omega t) / 2\pi \rho r^2. \qquad (5.11)$$

When deriving (5.11) it is assumed that the amplitude of the alternating component of current is considerably smaller than the direct component, i.e. $I_m \ll I_0$, and when deriving (5.10) $I_0 = 0$.

The equations (5.10) and (5.11) show that in the investigated magnetic field each cylindrical element of the fluid of unit mass is subjected to the effect of a force containing the stationary component F_{M0} and the non-stationary component F_{M1}:

$$F'_{M0} = -\frac{\mu_0 \chi I_m^2}{8\pi^2 r^3 \rho}; \qquad (5.12)$$

$$F''_{M0} = -\frac{\mu_0 M_s I_0}{2\pi r^2 \rho}; \qquad (5.13)$$

$$F'_{M1} = -\frac{\mu_0 \chi I_m^2 \cos 2\omega t}{8\pi^2 r^3 \rho}; \qquad (5.14)$$

$$F''_{M1} = -\frac{\mu_0 M_s I_m \cos \omega t}{2\pi r^2 \rho}. \qquad (5.15)$$

The effect of the stationary force on the fluid at fixed cylindrical boundaries may result in some distribution of static pressure. If the absolute value of this pressure is not very large, it has no significant effect on the nature of vibrational movement, and the components F'_{M0} and F''_{M0} can be generally excluded from further examination. The components F'_{M1} and F''_{M1} should be regarded as the driving force, and the minus sign may be rejected. This is equivalent to the change of the initial phase by π.

Further, to simplify the calculations, we examine the case of 'thin' cylindrical layers satisfying the following inequality:

$$h/r \ll 1, \qquad (5.16)$$

where $h \equiv r_1 - r_0$ is the thickness of the layer

Taking into account the monotonic form of the dependences F'_{M1} and F''_{M1} on r and the fact that the investigated range of variation of r is very small, the equations (5.14) and (5.15) can be replaced by their mean values in the range $r_1 - r_0$:

$$F'_{M1} = \frac{\mu_0 \chi I_m^2 (r_1 + r_0) \cos 2\omega t}{16\pi^2 \rho r_0^2 r_1^2} \tag{5.17}$$

$$F''_{M1} = \frac{\mu_0 M_S I_m \cos \omega t}{2\pi \rho r_0 r_1}. \tag{5.18}$$

Averaging of (5.14) and (5.15) enables us to use special functions in solving the differential equation (5.2).

Substituting successively (5.17) and (5.18) to the right-hand part of (5.2) gives two differential equations in general form:

$$\frac{\partial^2 u}{\partial t^2} = c^2 \left(r^{-1} \frac{\partial}{\partial r} r \cdot \frac{\partial u}{\partial r} \right) + A_v \cdot \cos \alpha_v t. \tag{5.19}$$

To implement the case 1 or 2, in equation (5.19) it should be accepted that

$$A_v = \frac{\mu_0 \chi I_m^2 (r_1 + r_0)}{(4\pi r_0 r_1)^2 \rho} \quad \text{and} \quad \alpha_v = 2\omega, \tag{5.20}$$

or

$$A_v = \frac{\mu_0 M_S I_m}{2\pi \rho r_0 r_1} \quad \text{and} \quad \alpha_v = 2\omega. \tag{5.21}$$

The differential equation (5.19) is solved by separating the variables r and t; for this purpose, $u(r, t)$ is written in the form

$$u = U(r) \cos \alpha_v t. \tag{5.22}$$

After substituting (5.22) into (5.19) and carrying out the simplest transformation, the new differential equation has the form

$$\frac{\partial^2 R}{\partial r^2} + \frac{1}{r} \frac{\partial R}{\partial r} + k^2 R = -\frac{A_v}{c^2}, \tag{5.23}$$

where $k = k_1 = 2\omega/c$ for case 1, and $k = k_2 = \omega/c$ for the case 2.

The differential equation (5.23) is non-homogeneous, and the solution of the appropriate homogeneous equation is well-known – it is expressed through the Bessel functions of the zeroth order

$$R^* = a_0 J_0(kr) + b_0 N_0(kr),\tag{5.24}$$

where J_0 is the Bessel function of the zeroth order; N_0 is the Neumann function of the zeroth order; a_0 and b_0 are arbitrary constant quantities.

To obtain the general solution of the non-homogeneous equation (5.23), the constant value A_v/c^2k^2 should be added to the solution (5.24):

$$R^* = a_0 J_0(kr) + b_0 N_0(kr) - A_v/c^2k^2.\tag{5.25}$$

As shown below, the highest value of the length of the resonance soundwave λ is $2h$. From this, on the basis of (5.16) $kr \gg 1$ and, consequently, the Bessel functions can be replaced by asymptotic equations:

$$J_0(kr) \approx \sqrt{\frac{2}{\pi kr}} \cdot \cos\left(kr - \frac{\pi}{4}\right);\tag{5.26}$$

$$N_0(kr) \approx \sqrt{\frac{2}{\pi kr}} \cdot \sin\left(kr - \frac{\pi}{4}\right),\tag{5.27}$$

and taking these equations into account the solution of (5.25) has the following form

$$R^*(r) = a_{01}\frac{\cos kr}{\sqrt{kr}} + b_{01}\frac{\sin kr}{\sqrt{kr}} - \frac{A_v}{c^2k^2}.\tag{5.28}$$

The values of the constants a_{01} and b_{01} are determined from the boundary conditions $R^*(r_0) = 0$ and $R^*(r_1) = 0$, because the restricting cylinders are regarded as absolutely rigid:

$$a_{01}\frac{\cos kr_0}{\sqrt{kr_0}} + b_{01}\frac{\sin kr_0}{\sqrt{kr_0}} - \frac{A_v}{c^2k^2} = 0;\tag{5.29}$$

$$a_{01}\frac{\cos kr_1}{\sqrt{kr_1}} + b_{01}\frac{\sin kr_1}{\sqrt{kr_1}} - \frac{A_v}{c^2k^2} = 0.\tag{5.30}$$

From the system of equations (5.29) and (5.30) we obtain:

$$a_{01} = \frac{A_v}{c^2 k^2} \frac{\sqrt{kr_0} \sin kr_1 - \sqrt{kr_1} \sin kr_0}{\sin k(r_1 - r_0)}; \qquad (5.31)$$

$$b_{01} = \frac{A_v}{c^2 k^2} \frac{\sqrt{kr_1} \cos kr_0 - \sqrt{kr_0} \cos kr_1}{\sin k(r_1 - r_0)}. \qquad (5.32)$$

Substituting a_{01} and b_{01} into equation (5.28) and then the resultant equation into (5.22) gives the solution of the differential equation (5.19) for the case 1 and 2:

$$u_1(r,t) = \frac{\mu_0 \chi I_m^2 (r_1 + r_0)}{\rho (8\pi\omega r_0 r_1)^2} \left[\frac{\sqrt{r_1} \sin k_1 (r - r_0) + \sqrt{r_0} \sin k_1 (r_1 - r)}{\sqrt{r} \cdot \sin k_1 h} - 1 \right] \cos 2\omega t$$

$$(5.33)$$

$$u_2(r,t) = \frac{\mu_0 M_S I_m}{2\pi\rho\omega^2 r_0 r_1} \left[\frac{\sqrt{r_1} \sin k_2 (r - r_0) + \sqrt{r_0} \sin k_2 (r_1 - r)}{\sqrt{r} \cdot \sin k_2 h} - 1 \right] \cos\omega t. \ (5.34)$$

Analysis of the equations (5.33) and (5.34), describing the movement of the fluid particles, confirms that the particles in the conditions of the model considered here carry out radial harmonic oscillations with a circular frequency 2ω in the case 1 and ω in the case 2. The amplitude of the oscillations depends on the magnetic parameters of the magnetic fluid – magnetic susceptibility χ in the case 1 and saturation magnetisation M_S in the case 2. If $\chi = 0$ or $M_S = 0$, which holds for conventional non-magnetic fluids, no oscillations appear. From a very large number of different fluid media, disregarding liquid metals, only the magnetic fluids are characterised by the unique property of the conversion of the energy of electromagnetic oscillations to the energy of elastic mechanical oscillations.

It may easily be seen that at $\omega = \omega_m = \pi c m / 2h$ (in case 1) and $\omega = \omega_m = \pi c m / h$ (in case 2), where $m = 1, 2, 3...$, the denominator of the equation in the square brackets of the equations (5.33) and (5.34) changes to 0 and the amplitude of oscillations has infinitely high values. Consequently, at alternating current frequencies of $\omega = \omega_m$ the oscillations become resonant. The formation of infinitely large oscillation amplitudes at resonance is the consequence of the assumption made regarding the absence of energy dissipation in the medium and the absence of emission of sound into the restricting cylinders.

We consider a specific example. Let it be that $h = 5$ mm at $r_0 = 50$ mm. Since at resonance $h = m\lambda/2$, the wavelength of the main resonance frequency will be 10 mm. Accepting for the magnetic fluid $c = 1300$ m/s, the main resonance frequency is $\nu = c/\lambda = 130$ kHz; this frequency is implemented in case 1 and case 2 by using alternating current with a frequency of 130 and 65 kHz.

Thus, the method can in principle be used for the direct excitation of the resonance ultrasound oscillations in the magnetic fluid. Later, the magnetic fluid emitter (MFE) with the same geometry of the magnetic field was investigated by P. Dubbelday (1980). He reported that the cylindrical model uses most efficiently the properties of the fluid magnetic material and concluded that it can also be used in the range of sonic frequencies of 0.1–3 kHz. Both fields – constant and exciting – have the azimuthal geometry and depend only on the radial distance, and the fields are generated by the currents, flowing through the conductors wound in the azimuthal direction.

5.3. The Q-factor of a magnetic fluid film – the emitter of elastic oscillations

The magnetic fluid source of elastic oscillations, functioning in the megahertz frequency range, was investigated for the first time by V.M. Polunin (1982) [22]. The emitter of elastic oscillations is a thin magnetic fluid film deposited on the surface of the solid. The theoretical model of this transducer of oscillations is an infinite plane-parallel magnetic fluid layer (Fig. 5.7).

The homogeneous viscous magnetic fluid with no electrical conductivity is situated on the surface of the solid made of an absolutely rigid, non-magnetic and non-conducting material and has the form of a plane-parallel layer with thickness h. The *XOZ* plane of the coordinate system coincides with the lower surface of the

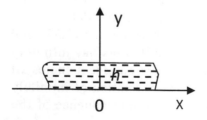

Fig. 5.7. The model of the emitter.

fluid layer. The driving force acting on the unit mass of the fluid along the y axis is:

$$F = f_0 \cos \omega t, \qquad (5.35)$$

where ω is the circular frequency, and the amplitude of the force f_0 is independent of the coordinates.

This equation for the driving force can be derived assuming, for example, that it is caused by the ponderomotive interaction of the fluid with the magnetic field:

$$H_x = H_0 + H_m(y) \cdot \cos\omega t, \; H_y = 0 \text{ and } Hz = 0, \qquad (5.36)$$

in which H_0 is independent of the coordinates, ∇H_m is a constant vector, $H_m \ll H_0$. Consequently

$$f_o = \mu_0 M |\nabla H_m| / \rho. \qquad (5.37)$$

The differential equation of oscillatory motion for this case has the following form:

$$\frac{\partial^2 u}{\partial t^2} = c^2 \frac{\partial^2 u}{\partial y^2} + \frac{\eta}{\rho} \cdot \frac{\partial^3 u}{\partial y^2 \partial t} + f_o \cdot \cos\omega t, \qquad (5.38)$$

where $\eta = \eta_v + 4\eta_s/3$ is the total viscosity of the fluid (η_s and η_v is its shear and volume viscosity).

For a stationary lower surface of the liquid layer and the free upper surface, the boundary conditions have the form

$$u\big|_{y=0} = 0 \text{ and } \frac{\partial u}{\partial y}\bigg|_{y=h} = 0. \qquad (5.39)$$

A system of flat standing waves forms in the liquid:

$$u = \sum_{m=1}^{\infty} u_{2m-1} \cdot \sin\frac{(2m-1)\pi y}{2h} \cdot \cos(\omega t + \psi_{2m-1}). \qquad (5.40)$$

Here u_{2m-1} is the amplitude of displacement of the fluid particles in the antinodes of the $2m-1$-th harmonics, and ψ_{2m-1} is the phase shift between these displacements and the driving force.

The parameters of the oscillatory motion u_{2m-1} and ψ_{2m-1} are to be determined. For this purpose, equation (5.40) is substituted into (5.38) and after transformations we obtain:

$$
\left[\sum_{m=1}^{\infty} \left(\omega_{2m-1}^2 - \omega^2 \right) u_{2m-1} \cdot \sin \frac{(2m-1)\pi y}{2h} \cdot \cos \psi_{2m-1} - \frac{\omega \eta}{\rho c^2} \sum_{m=1}^{\infty} \omega_{2m-1}^2 u_{2m-1} \times \right.
$$
$$
\left. \times \sin \frac{(2m-1)\pi y}{2h} \cdot \sin \psi_{2m-1} - f_o \right] \cdot \cos \omega t - \left\{ \sum_{m=1}^{\infty} u_{2m-1} \cdot \sin \frac{(2m-1)\pi y}{2h} \times \right.
$$
$$
\left. \times \left[\left(\omega_{2m-1}^2 - \omega^2 \right) \sin \psi_{2m-1} - \frac{\omega \eta}{\rho c^2} \omega_{2m-1}^2 \cdot \cos \psi_{2m-1} \right] \right\} \cdot \sin \omega t = 0,
$$

(5.41)

where $\omega_{2m-1} \equiv (2m-1)\pi c / 2h$.

Equality (5.41) should be fulfilled at any moment of time t and at any point y of the investigated material. Therefore

$$
\left(\omega_{2m-1}^2 - \omega^2 \right) \sin \psi_{2m-1} - \frac{\omega \eta}{\rho c^2} \omega_{2m-1}^2 \cos \psi_{2m-1} = 0, \tag{5.42}
$$

from which

$$
tg\, \psi_{2m-1} = \frac{\omega \eta \omega_{2m-1}^2}{\left(\omega^2 - \omega_{2m-1}^2 \right) \cdot \rho c^2}. \tag{5.43}
$$

In addition to this, we have

$$
\sum_{m=1}^{\infty} \left(\omega_{2m-1}^2 - \omega^2 \right) u_{2m-1} \cdot \sin \frac{(2m-1)\pi y}{2h} \cdot \cos \psi_{2m-1} - \frac{\omega \eta}{\rho c^2} \sum_{m=1}^{\infty} \omega_{2m-1}^2 u_{2m-1} \times
$$
$$
\times \sin \frac{(2m-1)\pi y}{2h} \cdot \sin \psi_{2m-1} - f_0 = 0. \tag{5.44}
$$

Multiplying the last term by the trigonometric series $\sum_{m=1}^{\infty} \dfrac{4 \sin\left[(2m-1)\pi y / 2h \right]}{\pi (2m-1)}$ converging to 1 in the range $0 < y < 2h$, we obtain

$$
\sum_{m=1}^{\infty} \left\{ u_{2m-1} \cdot \left[\left(\omega_{2m-1}^2 - \omega^2 \right) \cos \psi_{2m-1} + \frac{\omega \eta \omega_{2m-1}^2}{\rho c^2} \sin \psi_{2m-1} \right] - \right.
$$

$$-\frac{4f_0}{\pi(2m-1)}\Bigg\} \cdot \sin\frac{(2m-1)\pi y}{2h} = 0. \qquad (5.45)$$

The expression in the braces converts to 0. Replacing in this equation the functions $\cos \psi_{2m-1}$ and $\sin \psi_{2m-1}$ by $\mathrm{tg}\,\psi_{2m-1}$ using the well-known trigonometric identity and expression (5.43), we obtain

$$u_{2m-1} = \frac{4f_0}{\pi(2m-1)\sqrt{\left(\omega_{2m-1}^2 - \omega^2\right)^2 + \left(\dfrac{\omega\eta\omega_{2m-1}^2}{\rho c^2}\right)^2}}. \qquad (5.46)$$

Substituting (5.46) into (5.40) gives

$$u = \sum_{m=1}^{\infty} \frac{4f_0}{\pi(2m-1)\sqrt{\left(\omega_{2m-1}^2 - \omega^2\right)^2 + \left(\dfrac{\omega\eta\omega_{2m-1}^2}{\rho c^2}\right)^2}} \times$$

$$\times \sin\frac{(2m-1)\pi y}{2h} \cdot \cos\left(\omega t + \Psi_{2m-1}\right). \qquad (5.47)$$

The following expression of the resonance frequency ω_{2m-1}^2 can be written for each harmonics:

$$\omega_{2m-1}^r = \omega_{2m-1}\sqrt{1 - \frac{(\eta\omega_{2m-1})^2}{2(\rho c^2)^2}}. \qquad (5.48)$$

Because the second term in the radicand is small, equation (5.40) can be transformed to:

$$\omega_{2m-1}^r \approx \omega_{2m-1} = \pi(2m-1)c / 2h. \qquad (5.49)$$

The amplitude of the resonance frequency u_{2m-1}^r at not too high values of m is several orders of magnitude greater than the amplitude of the adjacent harmonics u_{2m+1} and u_{2m-3}. Actually, using (5.46) we obtain

$$u_{2m-3} / u_{2m-1}^r = \left[C_u \ell_1^2 (1 - \ell_1^2)^2 + \ell_1^6\right]^{-0.5} \qquad (5.50)$$

and

$$u_{2m+1} / u_{2m-1}^r = \left[C_u \ell_2^2 (\ell_2^2 - 1)^2 + \ell_2^6\right]^{-0.5}, \qquad (5.51)$$

where $C_u = \rho^2 c^4 / \eta^2 \omega^2_{2m-1}$; $\ell_1 \equiv (2m-3)/(2m-1)$; $\ell_2 \equiv (2m+3)/(2m-1)$.

The value η can be determined from the results of measurement of the coefficient of absorption of ultrasound α, linked with η by the dependence $\alpha = \omega^2 \eta / 2\rho c^3$. Using the results obtained in the previous chapter, it is assumed that $c = 1200$ m/s, $\alpha = 200$ m^{-1} at $= 25$ MHz. Consequently, for $m = 2$: $C_u \approx 10^5$, $\ell_1 = 1.3$, $\ell_2 = 7/3$, $u_1/u_3^r \approx 10^{-2}$, $u_5/u_2^r \approx 1.3 \cdot 10^{-4}$. The assumption on the smallness of the second term in the radicand of (5.48) is in agreement with the numerical value C_0.

For the frequencies close to the resonance frequency ω^2_{2m-1}, instead of the series (5.47) we can write:

$$u = \frac{4 f_0 \sin \dfrac{(2m-1)\pi y}{2h} \cos(\omega t + \psi_{2m-1})}{\pi(2m-1)\sqrt{\left(\omega^2_{2m-1} - \omega^2\right)^2 + \left(\dfrac{\omega \eta \omega^2_{2m-1}}{\rho c^2}\right)^2}}. \tag{5.52}$$

At $\omega = \omega^r_{2m-1}$:

$$u = \frac{4 f_0 \rho c^2 \sin \dfrac{(2m-1)\pi y}{2h} \cos(\omega t + (2m-1)\pi/2)}{\pi(2m-1)\eta \omega^3_{2m-1}}. \tag{5.53}$$

If there is only one ponderomotive mechanism of the excitation of oscillations that is acting, then at $\omega \approx \omega_{2m-1}$:

$$u = \frac{4 \mu_0 MGc^2 \sin \dfrac{(2m-1)\pi y}{2h} \cos(\omega t + \psi_{2m-1})}{\pi(2m-1)\rho \sqrt{\left(\omega^2_{2m-1} - \omega^2\right)^2 + \left(\dfrac{\omega \eta \omega^2_{2m-1}}{\rho c^2}\right)^2}}. \tag{5.54}$$

At $\omega = \omega_{2m-1}$:

$$u = \frac{4 \mu_0 MGc^2 \cdot \sin \dfrac{(2m-1)\pi y}{2h} \cdot \cos\left(\omega t + \pi/2\right)}{\pi(2m-1)\omega^3 \eta}. \tag{5.55}$$

The form of the equations (5.43) and (5.52) indicates that there is an analogy between the oscillatory movement, carried out by the fluid particles, and the mechanical system with concentrated parameters. This may be confirmed by introducing the notation

$\delta'_{2m-1} \equiv \eta\omega^2_{2m-1} / 2\rho c^2$ and regarding this quantity as the analogue of the damping coefficient of the mechanical system.

Therefore, the logarithmic damping coefficient Θ'_{2m-1} and the Q-factor Q'_{2m-1} of the transducer with only the internal losses taken into account, can be written in the form:

$$\Theta'_{2m-1} = \pi\eta\omega_{2m-1}/\rho c^2, \qquad (5.56)$$

$$Q'_{2m-1} = \rho c^2/\eta\omega_{2m-1} \qquad (5.57)$$

For these numerical values of α and c at $m = 1$ we obtain: $\Theta'_1 = 9.5 \cdot 10^{-3}$ and $Q'_1 = 330$.

The acoustic–mechanical analogy can be used for calculating the amplitude of oscillations of the particles as resonance u^r_{2m-1} taking into account both the internal losses and losses through emission.

It is well-known that

$$u^r_{2m-1} = u^s_{2m-1} Q_{2m-1} \qquad (5.58)$$

where u^s_{2m-1} is the effective value of the static displacement.
We have

$$u^s_{2m-1} = \frac{4f_0 \sin[(2m-1)\pi y / 2h]}{\pi(2m-1)\omega^2_{2m-1}}. \qquad (5.59)$$

To determine Q_{2m-1}, we use the property of the additivity of energy losses in the period ΔW: where $\Delta W = \Delta W_i + \Delta W_e$, where ΔW_i and ΔW_e are the internal losses and the losses through emission in the given period. If W is the mechanical energy of the system at time t, its logarithmic damping coefficient is $\Theta = -0.5|\Delta W|W$. Consequently,

$$\Theta_{2m-1} = \Theta'_{2m-1} + 0.5|\Delta W|W, \qquad (5.60)$$

If the inequality $\rho_c c_c \gg \rho c$ is satisfied (the wave resistance of the solid is considerably greater than the wave resistance of the fluid), a pressure antinode forms at the interface and a flat ultrasound waves with the following intensities passes through the interface

$$J = p^2/2\rho_c c_c \qquad (5.61)$$

where p is the amplitude of pressure at the antinode of the standing wave.

Ignoring the energy emited into the air, we obtain

$$|\Delta W_i| = \pi p^2 s / \rho_c c_c \omega, \qquad (5.62)$$

where s is the area of the active surface of the fluid layer.

The total energy of the liquid layer

$$W = (2m-1)\pi p^2 s / 8\rho c, \qquad (5.63)$$

Substituting (5.63) and (5.62) into (5.60) we obtain

$$\Theta_{2m-1} = \Theta'_{2m-1} + 4\rho c / (2m-1)\rho_c c_c, \qquad (5.64)$$

and since $Q_{2m-1} = \pi / \Theta_{2m-1}$, then

$$Q_{2m-1} = \frac{Q'_{2m-1}(2m-1)\pi \rho_c c_c / 4\rho c}{Q'_{2m-1} + (2m-1)\rho_c c_c / 4\rho c}. \qquad (5.65)$$

Let us assume that α and c have the values as previously, and $\rho = 1200$ kg/m^3, $\rho_c c_c = 133 \cdot 10^5$ kg/s $\cdot$ m^2 (the wave resistance of glass), then at $m \leq 3$ the inequality $Q'_{2m-1} \gg (2m-1) \rho_c c_c / 4\rho c$ is fulfilled and, consequently, we can write

$$\Theta_{2m-1} = 4\rho c / (2m-1) \rho_c c_c \text{ and } Q_{2m-1} = (2m-1) \pi \rho_c c_c / 4\rho c \qquad (5.66)$$

Thus, for $m = 1$ we obtain $Q_1 = 7.8$, which is almost identical with the Q-factor of quartz emitting into water. It is fully regular that for the absolutely rigid medium the expression (5.65) gives $Q_{2m-1} = Q'_{2m-1}$.

The Q-factor of the planar MFE in emitting into the solid medium is determined by the ratio of the acoustic resistance of this medium to the acoustic resistance of the transducer – the magnetic fluid, whereas for the solid vibrator, emitting into a liquid, the relationship is reversed.

At 'low' frequencies of the variation of the driving force, when $\omega \ll \omega_{2m-1}$, we can consider the oscillations of the 'thin' layer of the magnetic fluid. To estimate the amplitude of the oscillations of the particles in the 'thin' layer of the fluid, we can use the effective value of the static displacement u_{0ef}:

$$u_{0ef} = \frac{4f_o \sin(\pi y/2h)}{\pi \omega_1^2}. \qquad (5.67)$$

However, equation (5.67) does not take into account the contribution to the harmonics with the numbers higher than the first. The expression obtained from (5.47) is more accurate taking into account the fact that $\omega \ll \omega_1$, and the series $\sum\limits_{m=1}^{\infty} \cdot \dfrac{\sin[(2m-1)\pi y/2h]}{(2m-1)^3}$ converges in the range $0 \le y \le 2$ to the function $(\pi^3 y/16h)(1-y/2h)$.

This expression has the following form

$$u_{0ef} = \pi^2 f_0 y \ (1-y/2h)/4\omega_1^2 \ h. \qquad (5.68)$$

5.4. Excitation of sound in an unlimited magnetic fluid

The theoretical studies of the excitation of sound by a running magnetic field in an unlimited volume of a fluid were carried out by the Belarusian scientists V.G. Bashtovoy, B.M. Berkovskii and M.S. Krakov (1979).

The pressure in the magnetised fluid, situated in a heterogeneous magnetic field, is higher in the areas where the magnetic field is strong. In the stationary fluid the pressure distribution is determined by the equation $\nabla p = \mu_0 M \nabla H$. If the strength of the field is a periodic function of the spatial coordinates, a periodic distribution of pressure is formed in the fluid. If in addition to this, the field changes periodically with time, the pressure becomes a periodic function of both the spatial and time coordinates. In a compressible fluid, this results in compressions periodic in space and time which are nothing else but the forced sound waves.

The running magnetic field excites a running sound wave, and the standing magnetic field (proportional to $\cos kx \cos \omega t$) – a standing sound wave. It should be expected that if the characteristics of the driving force (and the role of the driving force in this case is played by the ponderomotive force $\mu_0 M \nabla H$) k and ω point coincide with the appropriate characteristics of the free sound waves, then the excitation of these waves by the magnetic field periodic in space and time will be resonant.

In fact, the frequency and phase of the free sound wave at every point coincide with the frequency and phase of the driving force and this creates suitable conditions for the supply to the system of the energy from the source of the external magnetic field.

The mathematical description of the investigated phenomenon is obtained in the simplest geometry of the infinite plane-parallel layer with thickness ℓ. The fluid fills the layer, and the running field is produced by the induction coils distributed at the boundaries of the layer. The OY axis is normal to the layer, as shown in Fig. 5.7. A special configuration of the magnetic fields is assumed:

$$B_x = \frac{B_0 shky}{shk\ell}\cos(kx - \omega t). \quad B_y = \frac{B_0 chky}{shk\ell}\sin(kx - \omega t)$$

$$H_x = \mu_0^{-1}B_x, \quad H_y = \mu_0^{-1}, \ B_y. \tag{5.69}$$

However, if the magnetic moment of the layer $\mathbf{M} \neq 0$, the configuration of the field is determined from a system of equations

$$\Delta\psi = \text{div}\mathbf{M}, \quad \Delta\mathbf{A} = -\mu_0 \ \text{rot}\mathbf{M}, \tag{5.70}$$

where $\mathbf{H} = -\nabla\psi$ and $\mathbf{B} = \text{rot}\mathbf{A}$. The system can be solved most easily in two cases. If $\mathbf{M} = \text{const}$, the running field coincides with (5.69) with the only difference that there is a constant addition. In the case of the linear dependence $\mathbf{M} = \chi\mathbf{H}$:

$$B_x = \frac{B_0 shky}{shk\ell}\cos(kx - \omega t), \ B_y = \frac{B_0 chky}{shk\ell}\sin(kx - \omega t), \tag{5.71}$$

$$H_x = \frac{B_x}{\mu_0(1+\chi)}, \quad H_y = \frac{B_y}{\mu_0(1+\chi)}, \quad M_x = \frac{\chi B_x}{\mu_0(1+\chi)} \text{ and } M_y = \frac{\chi B_y}{\mu_0(1+\chi)}. \tag{5.72}$$

Let it be that the speed of displacement of the fluid particles is v; ρ' and p' are the deviations from the equilibrium values of density ρ and pressure p, and in the linear approximation they are linked by the relation $p' = \frac{\partial p}{\partial \rho}\rho' = c_f^2 p'$, where c_f is the speed of sound in the magnetic fluid.

The fluid is regarded as non-conducting and the processes of viscous friction and thermal conductivity are ignored. The magnetic equation of state is assumed to be linear: $M = \chi H$. The process is assumed to be adiabatic.

Consequently, the system of ferrohydrodynamic equations has the form:

$$\rho\frac{\partial \mathbf{v}}{\partial t} = -c_f^2 \nabla\rho' + \frac{1}{2}\mu_0\chi\nabla H^2, \quad \frac{\partial \rho'}{\partial t} + \rho\,\text{div }\mathbf{v} = 0, \tag{5.73}$$

From the system of the equations we obtain in particular the expression for the oscillatory velocity:

$$v_x = \frac{(A_i/\rho)\cdot\omega/k}{(\omega/k)^2 - c_f^2}\cos 2(kx - \omega t), \quad v_y = 0, \qquad (5.74)$$

where $A_H = \mu_0\chi H_a^2/4$; $H_a = B_0/\mu_0(1 + \chi)\text{sh } k\ell$.

Thus, the running magnetic field excites longitudinal sound waves in the infinite layer of the magnetised fluid. The frequency of the excited sound is twice the frequency of the field. The condition for the formation of resonance is the agreement of the speed of the free sound waves and the speed of the running field.

Let it be that the running magnetic field has a constant component H^*, directed along the OY axis. Then

$$H^2 = H^{*2} + 2H^*H_a\, chky\sin(kx - \omega t) + H_a^2\left[sh^2ky + \sin^2(kx - \omega t)\right], (5.75)$$

$$v_x = \frac{(A_i/\rho)\cdot\omega/k}{(\omega/k)^2 - c_f^2}\cos 2(kx - \omega t) - \frac{4A_i}{\rho}\frac{H^*}{H_a}\frac{k}{\omega}chky\sin(kx - \omega t), \quad (5.76)$$

$$v_y = \frac{4A_i}{\rho}\frac{H^*}{H_a}\frac{k}{\omega}shky\cos(kx - \omega t). \qquad (5.77)$$

The expression for v_y (5.77) confirms that if the running field contains a constant component, not only the longitudinal but also transverse oscillations can be excited in the ideal fluid. Taking into account the viscous forces leads to the system of equations for $\mathbf{v}$ and ρ' in the form

$$\rho\frac{\partial\mathbf{v}}{\partial t} = -c_f^2\nabla\rho' + \frac{1}{2}\mu_0\chi\nabla H^2 + \eta\Delta\mathbf{v} + \left(\eta_v + \frac{\eta_s}{3}\right)\nabla(\nabla\mathbf{v}) \qquad (5.78)$$

$$\frac{\partial\rho'}{\partial t} + \rho\,\text{div }\mathbf{v} = 0. \qquad (5.79)$$

If again $H^2 = H_a^2\left[sh^2ky + \sin^2(kx - \omega t)\right]$, the longitudinal component of the speed of the wave is determined from the equation

$$v_x = v_{xa}\cos\left[2(kx - \omega t) + \varphi_{vx}\right], \qquad (5.80)$$

where φ_{vx} is the difference of the phases between the running magnetic field in the excited sound wave.

In this case

$$V_{xa} = \frac{\left(A_i/\rho\right)\cdot\omega/k}{\left\{\left[c_f^2-\left(\omega/k\right)^2\right]^2+4\omega^2b^2\right\}^{1/2}},\qquad(5.81)$$

$$tg\varphi_{vx} = \frac{2b\omega}{c_f^2-\left(\omega/k\right)^2},\qquad(5.82)$$

where $A_H = \mu_0\chi H_a^2/4$, b = $[\eta_v + (4/3)\eta_s]/\rho$.

As indicated by the equation (5.81), when the speed of the running magnetic field becomes close to the speed of sound in the fluid the amplitude of the excited sound oscillations increases to a certain extent – the resonance in the system with elastic energy dissipation.

It is characteristic that to excite the sound in the magnetic fluid it is not necessary to have a solid wall carrying out oscillations. It is sufficient to ensure that the running magnetic field has the appropriate parameters of the way vectors **k** and the circular frequency of oscillations ω.

5.5. The coefficient of ponderomotive elasticity of the magnetic fluid membrane

The equation for calculating the coefficient of ponderomotive elasticity of the magnetic fluid membrane k_p is derived using the model shown in Fig. 5.8 [24]. The centre of the mass of the magnetic fluid droplet, having the form of a disc (a column) with the radius R

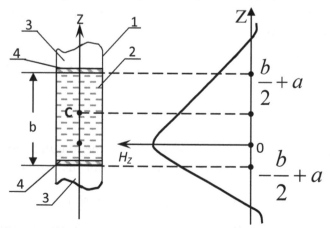

Fig. 5.8. The model for calculating k_p.

and thickness b, carries out small oscillations along the Z axis around the equilibrium position at the point $Z = 0$. The axial component of the force in the approximation of the 'weakly magnetic' medium is

$$\Delta f_z = 2\pi\mu_0 \int\limits_{-\frac{b}{2}+\Delta Z}^{\frac{b}{2}+\Delta Z} \int\limits_0^R \left[M_r \frac{\partial H_z}{\partial r} + M_z \frac{\partial H_z}{\partial z} \right] r \cdot dr \cdot dz, \qquad (5.83)$$

where M_r and M_z are respectively the radial and axial component of magnetisation of the fluid.

Taking into account in accordance with the topography of the magnetic field that $M_z \gg M_r$, because of the symmetry of the magnetic field in relation to the plane $Z = 0$ we obtain for $\Delta z \ll b$:

$$\Delta f_z = -2\mu_0 S \left(M_z \frac{\partial H_z}{\partial z} \right)_{z=-\frac{b}{2}} \Delta z. \qquad (5.84)$$

From this it follows that

$$k_p = 2\mu_0 S \left(M_z \frac{\partial H_z}{\partial z} \right)_{z=-\frac{b}{2}}. \qquad (5.85)$$

However, if the magnetic fluid is magnetised to saturation, then

$$k_p = 2\mu_0 S M_s \left(\frac{\partial H_z}{\partial z} \right)_{z=-\frac{b}{2}} \qquad (5.86)$$

It is assumed that under the conditions of this problem the surface tension forces can be ignored in comparison with the ponderomotive forces because of the small value of the capillary constant.

The frequency of oscillations is determined from the following equation:

$$\nu_m = \frac{1}{2\pi} \sqrt{\frac{2\mu_0 M_z}{\rho b} \cdot \frac{\partial H_z}{\partial z}}. \qquad (5.87)$$

If the magnetic fluid is not magnetised to saturation, then an additional perturbation of the magnetic pressure will form in the field normal to its surface associated with the discontinuity of the normal component of the strength of the magnetic field, and the equation (5.86) has the form:

$$k_p = 2\mu_0 S M_z \left(\frac{\partial H_z}{\partial z} + \frac{\partial M_z}{\partial z} \right)_{z=-\frac{b}{2}} ; \qquad (5.88)$$

$$v_m = \frac{1}{\pi} \sqrt{ \frac{\mu_0 M_z}{2\rho b} \cdot \left(\frac{\partial H_z}{\partial z} + \frac{\partial M_z}{\partial z} \right) }. \qquad (5.89)$$

5.6. Resonance frequency of oscillations of the magnetic fluid seal

In the magnetic fluid seals (MFS) and magnetic fluid gaskets (MFG), used widely in engineering [25], the magnetic fluid droplet overlaps the gap between the shaft and the sleeve as a result of the sustaining effect of the magnetic field concentrated in the area of the gap. Regarding the introduced model of ponderomotive elasticity as the first approximation, it will be used to estimate the resonance frequency v_p of the MFS. This is carried out using the expression for the critical pressure Δp_{cr} of the 'one-shank' seal:

$$\Delta p_{cr} = \mu_0 M_s (H_{max} - H_{min}), \qquad (5.90)$$

where H_{max} and H_{min} are the maximum and minimum strength of the magnetic field on the free surfaces of the magnetic fluid membrane.

Taking into account only the ponderomotive elasticity, calculated using equation (5.85), we obtain [26]:

$$v_r = \frac{1}{2\pi b} \sqrt{ \frac{2\Delta p_{cr}}{\rho} }. \qquad (5.91)$$

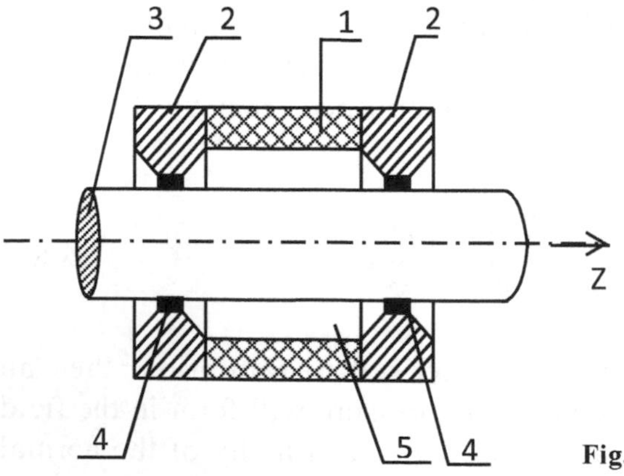

Fig. 5.9. Diagram of the MFS.

If $\Delta p_{cr} = 0.75 \cdot 10^5$ Pa, $b = 2$ mm, $\rho = 1.5 \cdot 10^3$ kg/m^3, then $v_r \approx 800$ Hz. In other words, the 'critical' working frequency of such a seal is equal to 800 Hz and, correspondingly, the angular speed of rotation of the shaft is ~5 $\cdot 10^3$ s^{-1}.

The seals with the symmetric distribution of the sealing elements are used in most cases. The design of the simplest MFS of such type is shown in Fig. 5.9.

The pole terminals 2, encircling the shaft of the magnetic material 3, are connected with the ring-shaped magnet 1. The magnetic fluid 4 is supplied into the gaps between the pole terminals and the shaft. The resultant closed cavity 5 is filled with air. This cavity is used as an elastic binding element between two identical magnetic fluid membranes.

Each magnetic fluid membrane is subjected to the effect of the force:

$$\rho S_r b \frac{d^2 Z_1}{dt^2} = k_g (Z_1 - Z_2) - k_p Z,$$

$$\rho S_r b \frac{d^2 Z_2}{dt^2} = k_g (Z_2 - Z_1) - k_p Z_p, \qquad (5.92)$$

where S_r is the area of the ring-shaped gaps; Z_1 and Z_2 are the displacements of the left and right membranes from the equilibrium position.

The system of equations (5.92) is the well-known system of two connected oscillators.

Such an oscillatory system has two normal frequencies:

$$\omega_1 = \sqrt{\frac{k_p}{\rho S_r b}} \quad \text{and} \quad \omega_2 = \sqrt{\frac{k_p + 2k_g}{\rho S_r b}}. \qquad (5.93)$$

The inequality $2k_g/k_p \ll 1$ determines the weak bonding condition and leads to:

$$V_0 \gg \frac{\rho_g c^2 S_r b}{2\Delta p_{cr}}. \qquad (5.94)$$

Assuming that $S_r = 5 \cdot 10^{-5}$ m^2, we obtain the restriction for the volume of the closed cavity: $V_0 \gtrsim 300$ mm^3.

If the inequality (5.94) and the initial conditions $Z_1 = Z_2 = 0$ and $Z' = 0$ are satisfied, the solutions of the system of equations (5.92) have the following form:

$$Z_1 \approx \frac{v_0}{\omega_1} \cos \Omega t \cdot \sin \omega_1 t,$$

$$Z_2 \approx \frac{v_0}{\omega_1} \sin \Omega t \cdot \cos \omega_1 t, \qquad (5.95)$$

where $\Omega \equiv k_g/(2\rho S_r b \omega_1)$.

Under these conditions, the magnetic fluid membranes carry out oscillations with a frequency ω_1, and the amplitude of these oscillations varies in accordance with the harmonic law with low frequency Ω, and this is accompanied by the periodic exchange of energy between them. If the frequency of the external periodic force determined by, for example, the eccentricity of the shaft, coincides with one of the normal frequencies (5.93), the resonance occurs. The amplitude of the oscillations in the investigated dissipativeless approximation increases without bounds.

The number of the unique properties of the magnetic fluid membranes is a prerequisite for using them in practice. For example, the effect of generation of the electromagnetic response – damping low-frequency electromagnetic pulses, formed immediately after the rapture of the magnetic fluid membrane, displaced from the region of the maximum magnetic field, can be used in the electroacoustics. Some chemical, physical–biological and pharmaceutical technologies use the processes of metered supply of gas into a reaction vessel. In this connection, it is interesting to use the magnetic fluid membranes as a valve capable of letting through specific portions of gas with the corresponding signalling in the form of acoustic and electromagnetic pulses. The magnetic fluid membranes are preferred in some situations as the main element of the pump – piston.

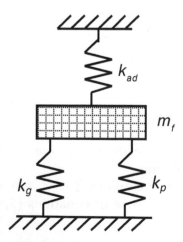

Fig. 5.10. The equivalent mechanical model of the oscillatory system.

5.7. Experimental method for determining the coefficient of ponderomotive elasticity

The coefficient of ponderomotive elasticity is measured using the method of the 'attached cavity' in which successive measurements are taken of the frequency of oscillations of the magnetic fluid membrane in a pipe open at one end v_1 and a pipe closed at both ends v_2 [26].

The equivalent mechanical model of the oscillatory system with the attached cavity is shown in Fig. 5.10. With the one end of the pipe open, the magnetic fluid membrane with the mass m_f is spring-loaded by the elasticity of the isolated gas cavity k_g and the ponderomotive type elasticity k_p.

When using the pipe closed at both ends, the elasticity of the attached gas cavity k_{ad} is added to these two elasticities. Consequently

$$v_1 = \frac{1}{2\pi}\sqrt{\frac{k_g + k_p}{m_f}}, \quad v_2 = \frac{1}{2\pi}\sqrt{\frac{k_g + k_p + k_{ad}}{m_f}}. \tag{5.96}$$

In the investigated case, the isolated chamber is part of the cylindrical pipe with the height h_0 and, therefore, the equation for the elasticity coefficient of the gas cavity (2.40) is used here in the form:

$$k_g = \frac{\gamma \pi d^2 p_0}{4h_0}, \tag{5.97}$$

where p_0 is the gas pressure in the cavity in the absence of oscillations; d is the pipe diameter; γ is the heat capacity ratio.

The formula for k_g can also be used; in this equation c is the speed of sound in air (2.41).

When solving the system of the equations with respect to k_p, we obtain

$$k_p = \frac{\pi^2 \rho_g c^2 d^4}{16 V_{ad}} \cdot \left[\frac{1}{n^2 - 1} - \frac{V_{ad}}{V_0}\right], \tag{5.98}$$

where V_{ad} is the volume of the attached cavity, $n \equiv v_2/v_1$.

Since the attached cavity is part of the pipe with a constant cross-section, the equation (5.98) has the following form:

$$k_p \cong \frac{\pi\,\rho_g c^2 d^2}{4h_{ad}}\left(\frac{1}{n^2-1}-\frac{h_{ad}}{h_0}\right).$$ (5.99)

The error of measurement of k_p using the method of the attached cavity is:

$$\frac{\Delta k_p}{k_p}=\frac{\Delta\rho_g}{\rho_g}+\frac{2\cdot\Delta c}{c}+\frac{2\cdot\Delta d}{d}+\frac{\Delta h_{ad}}{h_{ad}}+\frac{2n\cdot\Delta n}{n^2-1}.$$ (5.100)

The largest contribution to the error comes from the last two terms and their total value is in the range 10–15%.

5.8. A magnetic fluid chain with the ponderomotive type elasticity

Using the magnetic field, modulated in space, we can form a 'fluid chain' (FC) system. Here the FC are fluid droplets and the connected elements are the elastic gas cavities. Figure 5.11 shows the model of such a system in the form of a discontinuous fluid column. The chain of the magnetic fluid droplets is stabilised by a system of ring-shaped magnets placed coaxially with the pipe and spaced at distance d from each other. The system of the membranes is enclosed in an absolutely rigid cylindrical shell with a constant cross-section S; b

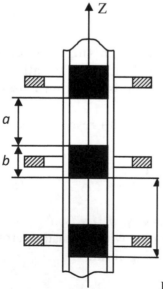

Fig. 5.11. The model of the magnetic fluid chain.

is the length of the fluid link; a is the thickness of the gas cavity; d is the identity period.

It is assumed that only the longitudinal sound wave (zero mode) propagates in the chain, there are no viscous friction and heat exchange processes, and the fluid is incompressible. In this approximation, this problem was solved by O.V. Lobova, V.M. Polunin and E.B. Postnikov in 2001 [27].

In the absence of heat exchange in the system the process of propagation of the elastic waves is adiabatic. The displacements of the fluid links from the equilibrium position are governed by the harmonic law and their values will be denoted in accordance with the number as U_{n-1}, U_n, U_{n+1}.

The elastic properties of the FC system are influenced by the gas and magnetoelastic components, determined by the interaction of the magnetic fluid with the source of the magnetic field. The displacement of the phase boundary U is caused by the parallel defect of both elasticity factors and, therefore, the coefficient of the quasielastic force of the system k is equal to the sum of the coefficients of the magnetoelastic and gas cavities: $k = k_p + k_g$.

It will be assumed that only the longitudinal sound wave (zero mode) propagates in the chain, the viscous friction and heat exchange processes are ignored, and the fluid is assumed to be incompressible.

In this case, the forces acting on the droplets with a number n from the side of the air cavities $\left(S\dfrac{p_0\gamma}{a}(U_{n+1} + U_{n-1} - 2U_n) \right)$ are added with the ponderomotive force acting on the droplet from the side of the magnetic fields during its displacement from the equilibrium position.

Under the condition of coincidence of the equilibrium position with the plane of symmetry of the magnet, this force is equal to:

$$f_m = -2\mu_0 SM_z\left(\frac{\partial H_z}{\partial z} + \frac{\partial M_z}{\partial z}\right)U_n. \tag{5.101}$$

As a result, the second Newton law for this droplet has the following form:

$$\rho_f Sb\frac{d^2 U_n}{dt^2} = S\frac{p_0\gamma}{a}(U_{n+1} + U_{n-1} - 2U_n) - 2\mu_0 SM_z\left(\frac{\partial H_z}{\partial z} + \frac{\partial M_z}{\partial z}\right)U_n. \tag{5.102}$$

Introducing $\chi' = \rho_g c^2 / \rho_f ab$, and also $\omega_m = \sqrt{\dfrac{2\mu_0 M_z}{\rho_f b}\left(\dfrac{\partial H_z}{\partial z} + \dfrac{\partial M_z}{\partial z}\right)}$ for the cyclic frequency of oscillations of the droplet under the effect of ponderomotive force, we write the equation describing the propagation of waves in the chain:

$$\frac{d^2 U_n}{dt^2} + \omega_m^2 U_n = \chi'\left(U_{n+1} + U_{n-1} - 2U_n\right). \qquad (5.103)$$

The equation (5.103) has the form of the standard equation of the connected interacting oscillators [28]. It is well-known that its solution has the form of the running wave:

$$U_n = A \cdot \exp i(\omega t - n k_w d), \qquad (5.104)$$

and the frequency and the wave number are linked by the dispersion equation:

$$\omega^2 = \omega_m^2 + 4\chi' \sin^2 \frac{k_w d}{2}. \qquad (5.105)$$

Analysis of the equation (5.105) leads to the conclusion that in the investigated system of excited waves with frequency ω only those waves propagate whose wavelength fits in the 'transparency range':

$$\omega_m \le \omega \le \sqrt{\omega_m^2 + 4\chi'}. \qquad (5.106)$$

This magnetic fluid chain operates as a band filter of sound oscillations. Only the perturbations with the frequencies from the 'window' (5.106) propagate over a sufficient distance, and others damp exponentially with increasing distance from the source. The effect of the oscillatory system is equivalent to the band LC filter.

Comparison of the elastic properties of the gas and magnetic subsystems is carried out on the basis of the parameter ψ:

$$\psi \equiv \frac{4\chi'}{\omega_m^2}. \qquad (5.107)$$

Assuming that the magnetic susceptibility of the magnetic fluid is equal to unity, we have

$$\psi = \frac{\rho_g c^2}{a\mu_0 MG},\qquad\qquad (5.108)$$

where ρ_g is the density of gas; c is the speed of sound in gas; M is the magnetisation of the fluid; G is the gradient of the strength of the magnetic field; μ_0 is the magnetic constant.

Taking into account the data obtained in this experiment for the field of the ring-shaped magnet, at $M = 20$ kA/m, $G = 4.5 \cdot 10^6$ A/m^2, $\rho_g = 1.29$ kg/m^3, $c = 340$ m/s, $a = 0.1$ mm, and using expression (5.108) we obtain $\psi \approx 13$.

When the value of a increases by an order of magnitude, the contributions of the magnetic and gas elasticity become close to each other. However, if $\psi \gg 1$, the role of the magnetic elasticity is small and the magnetic field is used only for maintaining the shape of the magnetic fluid droplet.

5.9. Rotational oscillations of a linear cluster in a magnetic field

The elastic properties of the nanodispersed medium – compressibility (section 2.3), the speed of propagation and the coefficient of absorption of the sound waves, may be characterised by the dependence on frequency, i.e., they are dispersed. The dispersion of these parameters is determined by the special feature of the structure of the substance and the type of its structure.

The processes of structure formation as a result of the dipole–dipole interaction take place in a specific sequence, with the structure of the largest particles first to form. These particles are characterised by large magnetic moments. The aggregates, consisting of the small particles, are less stable and easily fracture during rotation of the

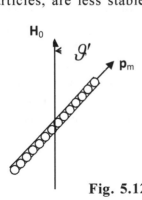

Fig. 5.12. Model of the linear cluster.

magnetic field. In this case, the dispersion of the elastic properties could be determined by the appearance in the magnetic colloid of aggregates consisting of ferroparticles of the fine fraction and having the resonance properties in the megahertz frequency range in the magnetic field [3]. The proposed resonance mechanism is associated with the forced rotational oscillations of the linear magnetic cluster around the direction of the external magnetic field $\mathbf{H}_0$ (Fig. 5.12).

The cluster is subjected to the effect of the rotational moment $\mathbf{M} = [\mathbf{p}_m \times \mathbf{B}]$ from the side of the magnetic field, and the magnitude of this moment is:

$$M_r = -\mu_0 M_S V H_0 \cdot \sin \vartheta', \tag{5.109}$$

where $M_S V = p_m$ is the magnetic moment of the chain; V is the volume of the chain.

The differential equation of the free non-damping oscillations in the approximation of small angles ϑ' is:

$$\ddot{\vartheta}' + \frac{\mu_0 M_S V H_0}{J} \vartheta' = 0, \tag{5.110}$$

where J is the moment of inertia of the chain in relation to the centre of rotation.

The resonance frequency of the oscillatory system ν_{res}:

$$\nu_{res} = \frac{\sqrt{12\mu_0}}{2\pi N_p d_p} \cdot \sqrt{\frac{M_S H_0}{\rho}}. \tag{5.111}$$

here N_p is the number of particles in the chain; d_p is the diameter of the particle with the shell of a surfactant.

The role of the driving force is played by the orientation mechanism proposed by Frenkel acting on the ellipsoidal particles in the ultrasound wave. Assuming that $H_0 = 100$ kA/m; $\bar{\rho} = 3 \cdot 10^3$ kg/m^3; $\varphi_m/\varphi_S < 0.6$; $M_S = 0.2 \, M_S'$ (taking the shell into account), $M_S' = 4.71 \cdot 10^5$ A/m; $d_p = 15$ nm, from equation (5.111) we obtain $\bar{\nu} = 15 \cdot 10^7/N_p$. This means that if the number of the particles is $N_p = 12$ then for $\nu \approx 6$ MHz the length of the chain is $\ell = 15 \cdot 12 = 180$ nm.

5.10. Oscillations of the form of the magnetic fluid droplet

The problem of the physical mechanism of electromagnetic excitation of the magnetic fluid active element, with its volume remaining unchanged, was discussed for the first time in studies by V.I. Drozdova, Yu.N. Skibin and V.V. Chekanov [29] in which theoretical and experimental investigations were carried out of low-frequency (2–3 Hz) axisymmetric oscillations of a spherical magnetic fluid droplet suspended in a non-magnetic fluid medium in a magnetic field. The theoretical model of elasticity of the oscillatory system proposed by the above authors takes into account the capillary forces and the forces of the ponderomotive effect of the magnetic field.

A magnetic fluid droplet, suspended in a non-magnetic fluid medium, has additional degrees of freedom associated with the deformation of the shape of the magnetised droplet. The process of oscillations is accompanied by perturbations of the internal magnetic field. When the droplet is ellipsoidal elongated in the direction of the external magnetic field, the demagnetising factor in the droplet decreases. This results in an increase of the strength of the magnetic field inside the droplet and, consequently, in even larger deformation of the droplet as a result of the increase of the pressure jump at the phase boundary and the poles of the ellipsoids (section 2.6). When the droplet is flattened along the axis, coinciding with the direction of the field, the situation is reversed. The demagnetising factor increases and the fields inside the droplet decreases which again increases the extent of deformation of the ellipsoids. Thus, in contrast to the capillary forces, always directed in the direction of restoration of the spherical shape, the ponderomotive forces of the magnetic field act in the opposite direction. Consequently, the elasticity of the oscillatory system, determined in the absence of the magnetic field by the surface tension forces of the fluid, decreases when the field is switched of, and the axial symmetry, directed along the magnetic field appears in deformation.

At $H = 0$ the frequencies of the free oscillations of the magnetic fluid droplet are:

$$\omega_o \big|_{H=0} = \sqrt{\frac{\sigma_o \ell(\ell-1)(\ell+2)(\ell+1)}{R^3 \left[\rho(\ell+1)+\rho_1 \ell\right]}}, \qquad (5.112)$$

where σ_0 is the surface tension coefficient; ρ and ρ_1 is the density of the magnetic fluid and of the non-magnetic liquid medium, respectively; $\ell - 1, 2, 3\ldots .$.

At $H \neq 0$ the frequencies of the intrinsic oscillations of the magnetic fluid droplet are determined from the following equation:

$$\omega_0 = \sqrt{\omega_0^2\Big|_{H=0} - \frac{\mu_0(\mu_i - \mu_e)H^2\ell(\ell+1)}{R^2\left[1 + (\mu_i/\mu_e - 1)N_x\right]\left[\rho(\ell+1) + \rho_1\ell\right]}}, \qquad (5.113)$$

where μ_i and μ_e are the magnetic susceptibilities of the substance inside the droplet of the magnetic fluid and outside it; N_x is the demagnetising factor.

The investigations were carried out on magnetic fluid droplets immersed in a water solution of calcium chloride having the density of the magnetic fluid. The removal of the droplets from the equilibrium position was carried out using a homogeneous magnetic field, generated by a system of Helmholtz coils. Under the effect of the field the droplets became ellipsoidal. After switching off the external magnetic field, the droplets carried out damping oscillations. The period of the free oscillations of the magnetic fluid droplet with a radius of 2.65 mm in a homogeneous magnetic field $H = 1.12$ kA/m was 0.24 s, whereas in the absence of the field it was 0.215 s.

In a later study of this subject, carried out by Yu.K. Bratukhin and A.V. Lebedev in 2002 [30], it was shown that the presence of viscosity decreases the resonance frequency of the oscillations of the droplet determined by the increase of the 'effective mass', i.e., by the appearance of attached mass.

5.11. Simple mechanism of volume magnetostriction

The group of the most probable mechanisms of excitation of the elastic oscillations of the magnetic fluid in an alternating magnetic field at the megahertz range frequencies includes the mechanism of

Fig. 5.13. Orientation of the stabiliser molecules: a) in the equilibrium state; b) in the non-equilibrium state.

volume magnetostriction – the mechanism of increasing the density of the medium in the vicinity of a ferroparticle during its rotational oscillations [3]. At rotational oscillations of the ferroparticles, the orientation of the elongated rod-shaped molecules of the shielding shells periodically changes (the number of these molecules according to the currently available data is ~10^3) and, consequently, the density of their molecular packing periodically changes. In the vicinity of the particle the fluid undergoes periodic tension and changes of the volume (Fig. 5.13), and this takes place synchronously in all the particles of the disperse phase.

If the period of the oscillations of the magnetic field is sufficiently small (shorter than the relaxation time of restoration of the equilibrium orientation of the molecules of the stabiliser), the rotational oscillations of the ferroparticles lead to changes of the volume of the fluid as a whole.

In a magnetising magnetic field $\mathbf{H}_0$ the RMS magnetic moment $\langle \mathbf{m}_* \rangle$ at the given temperature forms with the direction of the filed the angle $\langle \theta \rangle$. When applying a coaxial alternating field $H_m \ll H_0$, the value $\langle \theta \rangle$ changes in the range from $\langle \theta \rangle_{min}$ to $\langle \theta \rangle_{max}$. The magnetic moment $\langle \mathbf{m}_* \rangle$, 'frozen' into the ferroparticles, deviates in a single period of the oscillation of the field by some angle on both sides from the equilibrium direction. The rotation of the spherical particles in the viscous fluid–carrier causes oscillations of the volume of the fluid with double frequency.

As the equation of the magnetic state is non-linear, the deviation $\langle \mathbf{m}_* \rangle$ from the equilibrium direction will not be completely symmetric: it will be greater when decreasing the strength of the magnetic field and smaller when the strength is increased which in turn leads to the appearance of the harmonics of elastic oscillations with the frequency of the alternating field.

According to the results of investigation of acoustic double beam refraction and absorption of ultrasound waves in caster oil which consist mostly of rod-shaped molecules of rycin acid, the relaxation time of restoration of the equilibrium concentration is $\tau = 1.5 \cdot 10^{-7}$ s. The surfactant molecules in the shielding shell interact with the surface of the solid and consequently may be characterised by the longer time of structural rearrangement in comparison the above time and the duration of the Brownian rotational relaxation of the ferroparticles. Therefore, it may be expected that the proposed mechanism of excitation of the oscillations will be most effective at the frequency $v \geq 10^6$ Hz, i.e., in the megahertz frequency range.

Thus, the ponderomotive force is regarded as the elastic component and the driving force of the oscillatory system when the oscillations of the magnetic fluid element are accompanied by its flow with a constant volume (magnetic fluid membrane, magnetic fluid droplet). However, as can be seen, the electromagnetic excitation of the elastic oscillations of the magnetic fluid at the megahertz range frequencies can take place as a result of oscillations of the volume – volume magnetostriction of the fluid.

5.12. Magnetic levitation

The effect of the ponderomotive force is utilised in many systems: in magnetic fluid sealants, maintaining the pressure gradient of several atmospheres; in equipment for removing oil products from water surfaces; in magnetic heads of the loudspeakers with a magnetic fluid filling, improving their amplitude–frequency characteristics. The consequence of the effect of this force is the levitation effect.

In magnetic levitation, the non-magnetic body, placed in a magnetic fluid situated in a magnetic field with the gradient along the direction of the force of gravity, is subjected to the effect of additional ejecting force which may be many times greater than the weight of the displaced fluid [1, 11]. However, if the gradient of the strength of the magnetic field is directed vertically upwards, the magnetic levitation forces make the non-magnetic body heavier, prevent floating up, and ensure 'hovering' in a denser fluid medium. This phenomenon is the principle of the effect of separators of nonferrous metals.

The total force, which determines the condition of movement of the non-magnetic body in a magnetised magnetic fluid in the approximation of the weakly magnetic' medium can be presented in the form:

$$\vec{F} = (\rho_s - \rho)V\vec{g} - \mu_0 M V \nabla H, \qquad (5.114)$$

where ρ_s and V is the density and volume of the non-magnetic solids, respectively, M and ρ are the magnetisation and density of the magnetic fluid, H is the strength of the magnetic field, μ_0 is the magnetic constant.

The expression (5.114) gives the condition of buoyancy of the solid:

$$\rho_s < \rho + \mu_0 M \left| \nabla H \right| / g. \qquad (5.115)$$

The presence of the horizontal components in the gradient of the magnetic field determines the horizontal displacement of the body from the region with higher strength to the region with lower strength.

The air bubble, placed in a magnetic fluid, situated in the magnetic field with the strength gradient directed vertically upwards, is subjected to the effect of magnetic levitation forces which make the bubble 'heavier', prevent floating up and ensure 'hovering' in the liquid medium.

The total force which determines the condition of movement of the bubble in the magnetised magnetic fluid in the approximation of the 'weakly magnetic' medium can be written in the form [31]:

$$\vec{F} = -\rho V \vec{g} - \mu_0 M V \nabla H, \tag{5.116}$$

where V is the volume of the air bubble, M and ρ are the magnetisation and density of the magnetic fluid, H is the strength of the magnetic field, μ_0 is the magnetic constant.

Equation (5.116) gives the condition of levitation (hovering) of the air bubble

$$\mu_0 M \nabla H = -\rho \vec{g}, \tag{5.117}$$

If the viscous friction forces in the thin near-wall layer of the fluid are ignored and the condition $\mu_0 M |\nabla H| > |\rho g|$ is fulfilled, the magnetic levitation forces move the air bubble downwards.

The levitation effects which are very easy to produce in the magnetic fluid are used in the design of separation systems and density meters of non-magnetic materials, high-sensitivity tri-axial accelerometers, and a number of other advanced systems. This explains interest in exploring this phenomenon. The processes of magnetophoresis and Brownian diffusion influence the redistribution of pressure and levitation of the bodies in the magnetic fluid. Studies describing new systems operating on the basis of the magnetic levitation in the magnetic fluid have been published.

Questions for chapter 5

- *What is the magnetoacoustic effect in the magnetic fluid? What is its main mechanism?*
- *Describe briefly experimental setup for investigating the electromagnetic excitation of elastic oscillations in the magnetic fluid.*

- *What is the relationship between the frequency of the external alternating magnetic field and the frequency of oscillations of the particles in the cylindrical resonator?*
- *Describe method of determination of the logarithmic damping factor and the Q-factor of the magnetic fluid film – the emitter of elastic oscillations.*
- *Describe briefly the model used for determining the coefficient of ponderomotive elasticity of the magnetic fluid membrane.*
- *What are the practical applications of the chain of magnetic fluid droplets, stabilised by a system of ring-shaped magnets?*
- *What does the resonance frequency of oscillations of the linear magnetic cluster depend on?*
- *Explain the mechanism of deformation of the magnetic fluid droplet suspended in a non-magnetic medium.*
- *How does the mechanism of volume magnetostriction explain the excitation of ultrasound in a magnetic fluid in an alternating magnetic field?*
- *Write the equation for the total force acting on a non-magnetic body immersed in a magnetic fluid, taking the magnetic levitation effect into account.*

6

Comparison of equilibrium magnetisation of a nanodispersed magnetic fluid and a microdispersed ferrosuspension

Some experimental data for the equilibrium magnetisation of the magnetic fluid will be presented and compared with the appropriate data for ferromagnetic suspensions (FS) produced by the same procedure.

The force effect of a heterogeneous magnetic field on the FS is represented by the effect of remanent magnetisation. When the direction of the magnetic fields is changed to opposite in some strength range the specimen of the FS is pushed out from the magnetising solenoid.

In 1979, V.M. Polunin [32] investigated hysteresis loops for cyclic variation of H from 0 to 10 kA/m, the coercive force H_c and the remanent magnetisation M_r in the range of the strength of the external magnetic field from 10 to 75 kA/m, the dependence of remanent magnetisation on the holding time t, and also some possibilities of demagnetising the FS specimens. The investigated ferrosuspension was produced by thorough mixing of F-600 ferrite powder with castor oil. The volume concentration of the ferromagnetic was increased to 30%. The viscosity of the suspension, determined from the rate of discharge from a pipe, was (10 ± 0.5) Pa·s, i.e. an order of magnitude greater than the viscosity of castor oil. Taking into account non-Newtonian nature of the flow of the FS, the above results should be regarded as only an estimate of its static shear viscosity.

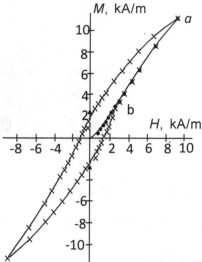

Fig. 6.1. The hysteresis loop of the ferrosuspension FS.

The disperse phase in the investigated FS sample was represented by magnetite particles of the single-domain size, and the disperse medium – by polyethylsiloxane fluid PES-2. The density and viscosity of the magnetic fluid were respectively 1.23 g/cm³ and 1 Pa·s.

Figure 6.1 shows the hysteresis loop of an FS. The curves were constructed taking the demagnetising factor into account. The magnetisation values, indicated by the full circles on the graph, were obtained in the process of initial magnetisation (branch 0–a), and those indicated by the crosses – in the process of cyclic variation of the strength of the field from 10 to −10 kA/m and vice versa. The black triangles indicate the values of remanent magnetisation, measured by the ballistic method. In the vicinity of $H = 0$ the error of the measurements greatly increases as a result of the increase of the error of recording $\Delta\Phi$.

In contrast to the classic hysteresis loop, typical of the solid ferromagnetics, the curve does not close at the point *a*, belonging to the tip of the loop, and closes at point *b* situated on the intermediate section of the initial magnetising branch. To explain this special feature, it is necessary to take into account the fact that the remagnetisation of the FS takes place mostly as a result of the rotation of the magnetised particles. The effect of this factor (together with the processes of intra-domain nature) in transition through the demagnetised state may result in a high curvature of the curve of the $M(H)$ dependence.

A second special feature of remagnetising the FS is the considerable (in this case ≈ 0.5 min) delay of establishment of M in relation to H in the vicinity of the point $H = \pm H_c$. This delay is also associated with the rotation of the ferrite particles. The reorientation time of the particles should depend strongly on the local viscosity which as a result of the dipole–dipole interaction between them may greatly exceed the viscosity of the disperse medium. This is one of the delay factors. The second factor is associated with the sequence of development of the process: the reorientation of the magnetic dipoles takes place mostly in the vicinity of the base of the measuring cylinder–vessel situated in the range of the highest strength of the field, and gradually propagates over the entire sample.

The graph can be used to obtain directly the values of the coercive force, magnetic susceptibility (determined as the ratio of the maximum values of M and H) χ in the investigated field strength range: $H_c = 1.3$ kA/m and $\chi = 1.2$.

Figure 6.2 shows the $M(H)$ dependence for the magnetic fluid for the cyclic variation of H in the range from 0 to 10 kA/m. The black circles indicate the direct course curve, corresponding to the increase of the strength modulus, and the crosses indicate the reverse course curve accompanied by a decrease of the strength from maximum modulus to 0.

The data presented in Fig. 6.2 show that within the measurement error range the direct and reversed course curves coincide with each other. The results, obtained on the basis of the ballistic method and the weighing method, indicate the complete absence of remanent magnetisation in the investigated sample of the magnetic fluid in the

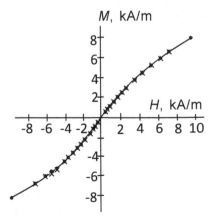

Fig. 6.2. Dependence $M(H)$ for the magnetic fluid.

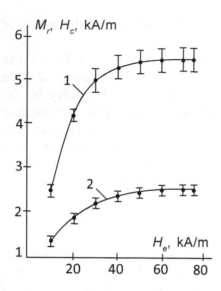

Fig. 6.3. Dependence of M_r and H_c on the strength of the field H_e.

range of the strength of the external magnetising field from 0 to 75 kA/m. In this respect, the magnetic fluid acts as an ideal magnetically soft material. Within the framework of the above considerations, the absence of any manifestations of magnetic hysteresis in the magnetic fluid in the static measurement conditions is explained by the short relaxation time of rotational diffusion of the magnetic moment of the magnetic particles of the single-domain dimensions suspended in the fluid and carrying out thermal Brownian motion.

According to the data in Fig. 6.2, the magnetic susceptibility of the investigated magnetic fluid is 0.86.

Figure 6.3 shows the dependence of remanent magnetisation M_r and coercive force H_c on the strength of the external magnetising field H_e for the FS. The dependences $M_r(H_e)$ and $H_c(H_e)$: 1 – M_r; 2 – H_c. The values of M_r were obtained by the ballistic method, the values of H_c by the weighing method. The magnetising exposure was 10 s. Prior to each subsequent magnetisation the sample was returned to the initial condition by slow withdrawal from the alternating magnetic field followed by mixing. The increase of H_e was accompanied by a gradually decreasing rate of the growth of both parameters, and the H_c/M_r ratio remained constant within the error range and equal to 0.5 which possibly indicates the existence of a physical relationship between them.

An important problem in the applied aspect is the problem of the magnetic ageing of the material based on a decrease of remanent magnetisation and variation of its main magnetic parameters with

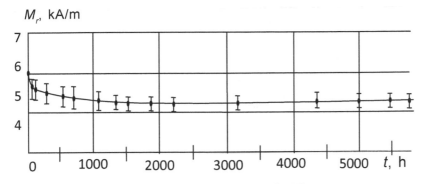

Fig. 6.4. $M_r(t)$ dependence for FS.

time. The variation of the remanent magnetisation of the FS sample M_r during holding is shown in Fig. 6.4.

During the experimental period (≈ 6000 s) M_r decreased by 16% from the initial value recorded 15 seconds after switching off the magnetising field. It is noteworthy that in the first 70–80 h M_r decreased by 8%, and by the same amount in the subsequent 1000 h. This was followed by stabilisation of magnetisation.

The time required for the passage of a spherical ferrite particle with a radius of 1.5 µm during its fall in castor oil through 1/3 height of the suspension (i.e., the duration of the complete delamination of the system) is ≈ 800 h. At the same time, after holding for 6000 h one can observe only a thin (≈ 0.5 mm) liquid phase film on the sample surface. This circumstance and also the relative stability of remanent magnetisation indicate the presence in the FS of a more or less continuous spatial structure, formed by the magnetised particles interacting over a distance.

The possibility of demagnetising the material is also of considerable practical interest. The well-known demagnetising method, based on the gradual withdrawal of the specimen from the alternating magnetic field with the strength of the field slightly higher than the strength of the magnetising field, reduces the magnetisation of the ferromagnetic suspension by at least 100 times. The demagnetising of this magnitude may also be achieved as a result of thorough mixing of the ferrosuspension. Naturally, the latter of the two possibilities of the magnetising is suitable only for liquid and paste-like systems.

The above results enable us to use the term 'liquid magnet' for the given FS sample because having fluidity, the sample (under the condition of constant shape) is capable of retaining its remanent

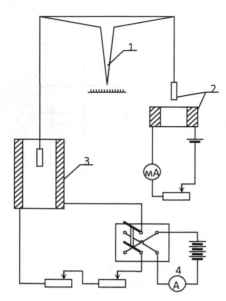

Fig. 6.5. The magnetic weighing method.

magnetisation of a long period of time at the level close to initial. A completely different behaviour in the process of static magnetisation is observed for a magnetic fluid sample which becomes completely demagnetised without any external effect of the field is switched off. This confirms the accuracy of the superparamagnetic model, used for magnetic colloids.

The measurements of magnetisation and coercive force in these investigations were carried out in a magnetic weighing equipment.

The magnetic weighing equipment is based on measuring the force acting on the investigated sample in a heterogeneous magnetic field. The magnetic weighing equipment, used in the present work, is shown schematically in Fig. 6.5.

The analytical damper balance 1 is designed for measuring the force acting on a cylindrical vessel with the investigated sample placed at the axis of the magnetising solenoid 3 in a section with a small strength gradient. The fixed position of the sample is obtained using the zero compensation electromagnetic device 2. The power source includes an accumulator, a set of rheostats, ammeter 4 and the current switch. The equation expressing the magnetisation of the investigated sample using the quantities obtained by direct measurements has the form:

$$M = gND^2lm/\Phi hId^2,\qquad(6.1)$$

where N and D is the number of turns and the diameter of the winding of the coil used for calibrating the strength gradient of the field; Φ is the variation of the magnetic flux penetrating the turns of the calibrating coil with length ℓ at the current intensity of the magnetising solenoid of 1 A; h and d is the height and internal diameter of the vessel; m is the weight difference of the specimens obtained at the current of the magnetising solenoid I; g is the freefall acceleration.

The measurement error by this method is $\Delta M/M = 5\%$.

Questions for chapter 6

*   *What is the difference between the magnetic fluid and the ferrosuspension?*
*   *What is the hysteresis loop in magnetising the ferrosuspension?*
*   *Explain the terms 'coercive force' and 'remanent magnetisation'.*
*   *What is the 'magnetically soft' and 'magnetically hard' material? Describe examples.*
*   *Which methods can be used to demagnetise a material?*
*   *Describe the principle of the magnetic weighing method.*

Rheological properties of suspensions

7.1. Newtonian and non-Newtonian fluids

Newton proposed the equation for the force acting on sheet S moving in a viscous fluid (gas):

$$F = \eta S \frac{du}{dz},\qquad(7.1)$$

where du/dz is the gradient of velocity in the direction z; η is the shear viscosity coefficient; S is the surface area to which the force F is applied (Fig. 7.1).

This equation is the mathematical form of writing the Newton law in hydro- and gas dynamics.

In addition to the shear viscosity coefficient, the kinematic viscosity coefficient is also considered. The kinematic viscosity

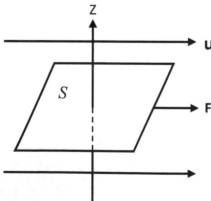

Fig. 7.1. Diagram of the problem.

coefficient is the ratio of shear viscosity to the density of the substance

$$v = \eta / \rho. \tag{7.2}$$

The shear viscosity coefficient η is described using equation (7.1) in the form of the ratio of the tangential stresses τ to the velocity gradient $\dot{\gamma}$, transverse to the direction of motion:

$$\eta = \tau / \dot{\gamma}. \tag{7.3}$$

To determine η it is necessary to carry out independent measurements of the quantities τ and $\dot{\gamma}$. If these quantities are varied and their ratio remains constant (i.e., if their dependence is directly proportional), such a fluid medium is referred to as *Newtonian*.

The concept of 'fluidity' is in no small measure associated with the specific rheological properties of the material, for example, the capacity for unlimited deformation under the effect of any small forces with the rate $\dot{\gamma}$, proportional to the applied stress τ:

$$\tau = \eta \, \dot{\gamma}. \tag{7.4}$$

The process of flow in the presence of external force using the model proposed by Ya.I. Frenkel will be examined in greater detail. In the presence of force F the potential relief forms an angle with the direction of the effect of the force. Since the work of the force along the path δ is equal to $F \cdot \delta$, the heights of the potential barriers on the right and left will be equal: $Q^{-} = Q - \dfrac{F\delta}{2} \hat{N}$ and $Q^{+} = Q + \dfrac{F\delta}{2} \hat{N}$ ($\hat{N}$ is the Avogadro number). Now the molecules will move more frequently from left to right than from right to left. In the absence of the force F the frequency of jumps to each side was equal to

$$v = \frac{2v_0}{z} e^{-Q/RT}, \tag{7.5}$$

where z is the coordination number (the molecule has z nearest neighbours).

Now, for the frequency of jumps from left to right we have:

$$v = \frac{2v_0}{z} e^{\frac{Q - \frac{F\delta}{2} \hat{N}}{RT}}. \tag{7.6}$$

For the frequency of jumps from right to left we obtain:

$$v = \frac{2v_0}{z} e^{-\frac{Q + \frac{F\delta}{2}\bar{N}}{RT}}. \tag{7.7}$$

The displacement of the particle during time dt is equal to:

$$\dot{u}dt = \delta(v_1 - v_2)dt. \tag{7.8}$$

The average velocity of movement of the particle is determined as follows:

$$\bar{u} = \frac{2v_0\delta}{z} e^{-\frac{Q}{RT}} \left(e^{\frac{F\delta}{2k_0T}} - e^{\frac{F\delta}{2k_0T}} \right). \tag{7.9}$$

Assuming $F\delta \ll k_0T$ and expanding the expression in brackets into a series, we obtain

$$\bar{u} = \frac{2\delta^2 v_0}{zk_0T} e^{-\frac{Q}{RT}} F. \tag{7.10}$$

In accordance with the definition of shear viscosity:

$$\eta = F/(S\frac{\Delta\dot{u}}{\Delta z}) = F/\delta\bar{u}. \tag{7.11}$$

With these assumptions, the equation for shear viscosity has the form:

$$\eta = \frac{zRT}{2\delta^3 v_0} e^{\frac{Q}{RT}}. \tag{7.12}$$

The expression for shear viscosity (7.12) derived on the basis of modelling the representations for many fluids is in good agreement with the experimental data. The characteristic feature is the absence of any dependence on the rate regime of the hydrodynamic process. This special feature is characteristic of Newtonian fluids.

However, the relationship (7.4) can greatly change its form in dependence on the deformation regime, the strength of the magnetic field, kinematic and dynamic characteristics of the flow. These media are referred to as *non-Newtonian* media.

Outsider the field, the low-concentration magnetorheological suspensions with a low viscosity carrier medium behave as Newtonian fluids whose viscosity does not depend on the shear rate, but increases with increase of the solid phase concentration in accordance with the Einstein equation:

$$\eta = \eta_0(1 + \alpha\varphi), \tag{7.13}$$

where η_0 is the viscosity of the carrier medium, φ is the volume concentration of the particles of the disperse phase, α is a dimensionless number ('intrinsic viscosity') which depends on the shape of the particles.

The non-Newtonian fluids include the so-called linear–viscoplastic media characterised by the additivity of the tangential stresses determined by the effect of plasticity and viscosity. Such viscoplastic media are described by the Bingham–Shvedov equation:

$$\tau = \tau_0 + \eta_p \dot{\gamma}, \tag{7.14}$$

where τ_0 is the shear stress, η_p is plastic viscosity.

The viscosity of the concentrated suspensions is described by the dependence derived by Vand [49]:

$$\eta/\eta_0 = \exp\,[(2.5\varphi_g + 2.7\varphi_g^2)/(1-0.609\varphi_g^2)].$$

In a study published in 1991, V.M. Buzmakov and A.F. Pshenichnikov presented an equation for the relative viscosity of a solution of the 'magnetite in kerosene' type:

$$\eta/\eta_0 = 1 + 4.4\varphi_g + 28.5\varphi_{2g},$$

derived by approximating the experimental data for temperatures of 25, 40 and 60°C.

The most probable reason for the high viscosity of the ferrocolloids is the presence the aggregates in the fluid. At a low shear rate the behaviour of the aggregated magnetic fluid becomes non-Newtonian and its deviation from the Newtonian behaviour becomes greater with a decrease of the shear rate and increase of the volume fraction of the solid phase.

In magnetic fluids, in addition to the hydrodynamic interactions, there is also the magnetic interaction of the particles influencing their relative motion and, therefore, the viscosity of the magnetic fluids also depends on the strength of this interaction. In a stable magnetic fluid the magnetic interaction can be ignored. In this case, the viscosity of the magnetic fluid is determined by the hydrodynamic concentration of the particles $\varphi_g = p\varphi$ and corresponds to the relationships obtained for the suspensions of non-magnetic particles. Here p is a coefficient which is independent of the concentration of the solid phase.

The viscosity of the real magnetic fluids may depend on the prior history of the sample (i.e., on preliminary external influences, such as mixing and magnetising) and the shear rate.

The aggregation of the magnetic fluid is accompanied by the increase of the effective hydrodynamic concentration φ_g. In addition to this, coarse-grained structures may penetrate through the entire volume of the magnetic fluid and be inhibited by its boundaries. Both mechanisms increase the viscosity and non-linear dependence of viscous stresses on the strain rate.

The viscosity and rheological behaviour of the magnetic fluids are influenced by the variation of temperature. Most of all, the temperature influences the viscosity of the carrier medium of the magnetic fluid, surfactants and, in addition to this, the variation of temperature influences the contribution of rotational diffusion to viscosity and the process of aggregation of the particles in the fluid. Therefore, the temperature dependences of the viscosity of the magnetic fluid and the carrier medium differ. This difference becomes larger with an increase of the temperature of the magnetic phase in the magnetic fluid and with increasing temperature. Viscometric experiments have confirmed that there is a large difference of the temperature dependence of the effective viscosity of the magnetic fluid from that for the carrier medium.

7.2. Magnetorheological effect

The magnetorheological effect is the significant change of the mechanical properties (viscosity, plasticity, elasticity) of some suspensions under the effect of magnetic fields.

Large changes are also observed in thermal and electrical conductivity and magnetic permittivity. The magnetorheological effect offers considerable possibilities for the direct effect of electrical signals in controlling hydrodynamic, heat and mass transfer, electrical and magnetic properties of the liquid fluid media.

The viscosity of a magnetorheological suspension (MRS) may increase by up to 100 times with an increase of the strength of the magnetic field (in contrast to magnetic fluids where viscosity increases by 10–30%).

The magnetorheological suspension may consist of a ferromagnetic disperse phase in the form of highly dispersed powders of iron, for example, carbonyl iron with average size of the particles in the range from 2 to 10 μm. Powders of cobalt, nickel or their mixtures with iron powders may also be used. In contrast to the magnetic fluids, the volume fraction of the ferromagnetic in the magnetorheological suspension is 10–50% and this results in the occurrence of the magnetorheological effect based on the large change of viscosity when a magnetic field is applied. The viscosity of some of the suspensions may change by several orders of magnitude.

The magnetorheological suspensions are characterised by high fluidity, controlled by the magnetic field, change their viscosity depending on the strength and gradient of the magnetic field, are characterised by the autolevitation effect, i.e., displace non-magnetic solids from their volume or suspend magnetic solids in their volume. These systems are characterised by the hysteresis of structure formation in the magnetic field. After switching off the magnetic field the existing structure is retained to a certain extent and thorough mixing is required to remove this structure.

The hysteresis is caused by the irreversibility of the magnetising processes. The magnetic characteristic of this phenomenon is the hysteresis loop – the dependence of magnetic induction B (magnetisation M) on the strength of the magnetic field H in cyclic remagnetising of the sample. The main parameters are: remanent induction B_r, remaining in the specimen after removing the external field; coercive force H_c – the demagnetising field with the opposite direction which must be applied to the specimen in order to remove the remanent induction; the area of the loop characterising the energy losses through hysteresis in a single remagnetising act (Fig. 6.1).

It is well-known that in heating the spontaneous magnetisation decreases and at the Curie point rapidly decreases to 0, – the magnetics transform to the paramagnetic state in which the magnetic moments remaining in the substance (electronic orbits and spins) are slightly oriented by the field because the effect of the field is inhibited by the thermal motion.

The existence of these temperatures is caused by the fact that in sufficiently intensive heating the disordering effect of the thermal motion on the orientation of the magnetic axis of the atoms becomes so large that it overcomes the forces of interaction of the atoms ensuring spontaneous magnetisation of the domain.

The strongest effects, accompanying the magnetising of the disperse ferromagnetics, are observed in the suspension of the magnetically soft materials, in particular, in the suspensions of carbonyl iron or carbonyl nickel. The carbonyl metals are produced on the industrial scale and can be used efficiently for utilising all the possibilities associated with the effects of interaction of the particles in the magnetic field. Other types of magnetically soft materials, including ferrites, are also available.

The suspensions acquire distinctive plasticity in the magnetic field with a high yield stress (shear strength). For the chain structure and the magnetic nature of the bonding forces of the particles the shear strength is equal to the specific (per unit volume) energy of the magnetic interaction of the particles.

At low shear rates the shear resistance is almost completely independent of the strain rate so that it is possible to transfer the torque which is almost completely independent of the sliding speed of the friction couplings. This transfer characteristic can also be useful in limiting and magnetically controlled couplings, brakes and other devices. The conditions for application of these devices may be very stringent when using suspensions based on metallic low-melting media.

The concentrated suspensions of the magnetically hard ferrites are characterised by a high limiting yield stress also in the absence of the external magnetic field because the particles of the suspensions are spontaneously magnetised to saturation. Of interest for practice is the possibility of the directional synchronous rotation of the particles in the alternating magnetic field in these suspensions.

In shear deformation of the suspension the direction of rotation coincides with the shear direction and this is reflected in a large decrease of the viscosity of the suspension and even in the effect which can be formally described as the negative viscosity. In reality, this means that the suspension starts to flow under the effect of the field even in the absence of any external force causing the flow. Because the suspension 'does not know' in the absence of this force the direction of the flow, the flow may have different sometimes surprising forms. In the movement from the wall of a vessel to its centre the flow takes place simultaneously in two opposite directions. The presence of the flow is reflected in the fact that ridges of moving waves form regularly on the surface of the suspension. The suspension is rapidly heated as a result of continued rotation of the particles in the viscous medium and, consequently, can be used as a distributed heat-generating agent.

7.3. Physical nature of the magnetorheological effect

7.3.1. Role of the structure formation of ferrosuspensions in the formation of magnetic susceptibility

The process of clusterisation of the magnetic system takes place in the magnetic field: magnetic particles form chains stretched along the field. At a sufficiently high level of the general dipole or molecular interaction the particles form chains, clusters with a flux closed inside them or a continuous spatial network even in the absence of the external field. In the suspension of single-domain particles this effect is quite strong: in mixing, the particles merge into spherical granules with the same sizes and the diameter of up to several millimetres. Mechanical mixing has an effect similar to that of heating of the suspension, i.e., causes 'swinging' of the particles in the clusters and the rearrangement of the structure. The clusters are characterised by higher density because the increase of the density of the clusters is determined by the dipole interaction of the particles and has advantageous energy characteristics.

The molecular bonding forces do not depend on the orientation of the magnetic dipoles and cause random bonding of the particles into porous aggregates. If these forces are considerably higher than the dipole forces, the hydrodynamic forces, required for loosening the cluster, become also higher than the dipole forces

In the clusters, the structural network and the granules the local field has a random direction, and the orientation of the magnetic moments of the particles coincides on average with the direction of the local field. In this case, the local field prevents the orientation of the particles along the external field and reduces the magnetic susceptibility of the suspension.

If instead of clusters there are chains of N particles, their behaviour in the field is identical with that of a particle with the moment Nm_* and, therefore, the initial susceptibility increases N times: $\chi = \chi_0 N$. In this case, the susceptibility is determined only by the concentration of the magnetic phase φ and the structure of the suspension.

However, if the increment of the magnetisation would be caused only by the spontaneous formation of the chains, the initial section of the graph $M(H)$ would be linear because in this case the initial susceptibility in the weak fields is constant. In the aggregation, forced by the field, N increases with increase of the strength of the field as a result of including larger and larger numbers of particles in the aggregates and decreasing the average distance between the aggregates r_0 for the particles.

The homogeneity of the diluted ferrosuspensions can be maintained constant only by mixing, which fractures large flakes to finer fragments. This decreases the average number of the nearest neighbours z and increases the initial magnetic susceptibility. The interruption of mixing decreases the initial susceptibility.

At the constant intensity of mixing the specific susceptibility decreases with increasing particle concentration because the size of the flakes and the average number x of the neighbours in interaction increase. In the $BaO \cdot 6Fe_2O_3$ suspensions with the particles approximately 1 µm in size and in suspensions $\gamma \cdot Fe_2O_3$ with the particles approximately 0.3 µm in size the value φ/χ, which is the inverse specific susceptibility, approaches 1.5, i.e. z approaches 2.

The concentrated suspensions of the single-domain particles ($\varphi > 0.1$) retain their homogeneity (do not separate) also without mixing as a result of the formation of a continuous network of the inter-connected particles. At the strength of the field smaller than the coercive force of the particles, the suspensions are magnetised mainly as a result of rotation of the particles. Rotation can take place if the bonding forces of the particles are sufficiently small. These forces weaken in the formation of adsorbed layers on the particles and, therefore, the magnetisation is strongly affected by the addition of surfactants to the liquid medium, for example, oleic acid to the particles of $BaO \cdot 6Fe_2O_3$ in the hydrocarbon medium. The addition of a small amount of a surfactant makes the magnetisation curve to have the form typical of the limiting hysteresis loop. After hardening the medium of the same suspension the magnetisation of the produced system is determined by the rotation of the magnetisation vector inside the particle.

7.3.2. Nature of non-Newtonian viscosity in a ferrosuspension

The dispersed nature of the magnetic component results in the dependence of viscosity η on the magnetisation of the liquid M and anisotropy of the particles. This dependence (rotational viscosity) is explained and analysed as a result of the interaction of the individual particles with the external field and the medium. At the same time, the rheology of the magnetic fluid and the ferrosuspension is related to the interaction of the particles with each other which, in particular, is reflected in the formation of chains.

In the group of the physical models of the disperse media it is important to mention three greatly different models: the Einstein model for a suspension of non-interacting particles, the Frenkel–Eyring model for highly concentrated suspensions with a crystal-like order in the distribution of the particles [33], and a chain model of the moderately concentrated disperse systems.

The magnetic fluid with the maximum possible concentration may form when the fluids are used in different devices with a strong and an inhomogeneous magnetic field. In the concentrated disperse systems the flow is a consequence of jumps of the particles into the adjacent vacant node of the crystal-like lattice. The stresses, generated in the lattice by the external force, change the magnitude of the potential barrier of 'permeation' of the particles to the vacant anode of the lattice. This change of the barrier creates preferential migration of the vacancies in one direction and of the particles in another.

Special attention will be paid to the chain model of the moderately concentrated disperse systems with inter-connected particles, developed by E.E. Bibik [34].

In the magnetic field the particles merge into chains and chains are aligned in the field direction. The flow is normal to the field – the simple shear flow. In this case, the chains fracture by hydrodynamic forces and are restored as a result of the dipole interaction of the particles to some hydrodynamically equilibrium length $l = Nr_0$, where N is the number of particles in the chain, and r_0 is the distance between the adjacent particles of the chain. The thermal breakdown of the chains can be taken into account postulating the existence of the thermodynamically equilibrium length $l_0 = N_0 r_0$. The latter is regarded as independent of $\dot{\gamma}$ at sufficiently low rates. To make sure that the model of the chains is similar to an actual system in which the chains can move in the free liquid matrix, it is necessary to introduce another geometrical parameter of the problem of the flow in a suspension of chains represented by the width of the gap (channel) h.

In the magnetic field the chain of particles most as a single unit retaining the orientation along the field. The particles, distributed on different sides from the centre of the chain at the distance $\pm x \leq l/2$, are bypassed by the medium in the opposite directions and, therefore, are also subjected to the effect of the friction force in opposite sides, with the friction force equal to $\pm\beta\dot{\gamma}x$, where β is the coefficient of resistance of the particle, for example, $6\pi\eta_0\alpha$ for spheres with radius α. Summing up one half of the chain with respect to the particles, these forces create in its centre the shear force $\beta\dot{\gamma}r_0\,(N^2-1)/8$. At the hydrodynamically equilibrium size of the chain this force is equalised by the bonding force of the particles. Therefore, the equation of the structural state of the magnetic fluid, determining the hydrodynamically equilibrium size of the chains, has the form

$$\beta\dot{\gamma}r_0(N^2-1)/8 = 3\mu_0\overline{m}_*^2/r_0^4, \tag{7.15}$$

and at $N \gg 1$

$$\beta\dot{\gamma}l^2 = 12U_c, \tag{7.16}$$

where $U_c = 2\mu_0\overline{m}^2/r_0^3$ is the binding energy of the adjacent particles of the chain as a result of the dipole–dipole interaction.

The resistance to shear deformation is written in the form

$$\tau = [\frac{2}{3}\beta n r_0^2 \frac{N^2-1}{8} + \eta_0(1+p\varphi)]\dot{\gamma}, \tag{7.17}$$

where n is the concentration of the particles, p is a coefficient of the order of unity which depends on the strength of the field H.

At low shear rates at which l is independent of $\dot{\gamma}$ and equal to l_0, the shear stress, according to (7.18), is proportional to the shear rate. The first term in the square brackets, representing the structural part of the viscosity of the magnetic fluid, is constant and at $N > 1$ is equal to $\beta n l_0^2/12$. The expression for the structural viscosity η_c can be easily generalised to the case of polydispersed chains because the contribution of the chains with the length l (or the number of particles N) is proportional to their concentration n_l with the weight $(N^2 - 1)$:

$$\eta_c = \frac{\beta}{12}\sum_{\upsilon=1}^{\infty} n_l(N^2-1). \tag{7.18}$$

In accordance with (7.18), the structural viscosity in this case depends on the strength of the field because the field influences the distribution function of the particles in chains of different lengths.

If, regardless of the factors which determine the equilibrium length of the chains (the thermal or hydrodynamic breakdown of the chains), it appears that the equilibrium length is greater than the width of the gap, the actual length of the chain will be equal to the width of this gap along the direction of the field. The equation of the structural state has in this case a very simple form:

$$N = h/r_0, \tag{7.19}$$

and the structural viscosity, as in the previous case, does not depend on $\dot{\gamma}$ but becomes dependent on the width of the gap h:

$$\eta_c = \beta n h^2/12, \tag{7.20}$$

It is noteworthy that in this flow mode (creep mode) the viscosity determined by the strength of the field and is not a characteristic of the fluid invariant in relation to the equipment (depends on h).

At a high level of the dipole interaction of the particles the creep modes can also be observed at high shear rates. According to (7.17), the potential length of the chains is greater than the length of the gap, and the actual length is equal the width of the gap up to the rates $\dot{\gamma} < \dot{\gamma}_1$, where

$$\dot{\gamma}_1 = 12 U_c / \beta h^2. \tag{7.21}$$

At higher rates l is determined by the equations (7.15) and (7.16). Using them in (7.17), we obtain a different flow law:

$$\tau = \tau_c + \eta^* \dot{\gamma}, \tag{7.22}$$

where $\tau_c = n U_c$ is the structural part of shear resistance; $\eta^* = \eta_0(1 + p\varphi)$ is the plastic viscosity identical with the Einstein viscosity.

The resultant rheological equation expresses the well-known empirical law of the flow of plastic materials (Shvedov–Bingham law). This also explains its nature: the plastic flow of the disperse systems – as a result of the existence of the hydrodynamically equilibrium state of the structure of the system, and this structure is not necessarily a three-dimensional frame of bonded particles.

The last equations show that the general viscosity of the magnetic fluid or ferrosuspension $\eta = \tau/\dot\gamma$ in the plastic flow mode decreases with increasing flow speed in accordance with the law $\eta = \eta' + \tau_c$ $\dot\gamma$ or $\eta = \eta^* \tau/(\tau - \tau_c)$. This dependence should remain unchanged up to the fracture of the chains into individual particles which, according to (7.15), starts at the shear rate of the order of $12U_c/\beta r_0^2$. The equation of the structural state becomes trivial in this case ($N = 1$) but together with the equation (7.17) leads to an inaccurate result – the structural viscosity at $\dot\gamma > 12U_c/\beta r_0^2$ is equal to 0, i.e., it leads to the rheological equation $\tau = \eta^*\dot\gamma$. After fracture of the chains to individual particles of the chain model cannot be used any longer – the state of the structure cannot be described by the number of particles N in the chain.

The nature of transition from the creep mode to the plastic flow with increasing shear rate depends on the value of the ratio l_0/h of the thermally equilibrium size of the aggregates to the width of the gap. If $l_0/h < 1$, the transition is 'stretched' over a relatively wide range of rates – from $\dot\gamma_T$ to $\dot\gamma_r$. The lower boundary of this range coincides with the highest shear rate at which the effect of the hydrodynamic forces on the chain is weaker in comparison with the thermal effect (breakdown). The upper boundary is the shear rate at which the ratio between the above-mentioned factors disrupting the chain is reversed – the thermal effect is small in comparison with the hydrodynamic effect.

In the rotating field the chains follow the field and transfer the torque to the vessel with the magnetic fluid through a viscous medium. The specific (per unit volume of the magnetic fluid) value of the torque M_r follows the same dependence on ω (at $\omega < \omega_k$, i.e., at the rotational frequencies of the magnetic field at which chains exists as a result of the dipole–dipole interaction of the magnetic particles), as the dependence of τ on $\dot\gamma$:

$$M_r = \left(\frac{2}{3}\beta n r_0^2 \frac{N^2-1}{8} + p\eta_0\varphi \right)\omega. \qquad (7.24)$$

In contrast to equation (7.18) in this equation there is no term $\eta_0\omega$ associated with the macroscopic motion of the medium itself.

The formation of the chains in the magnetic field is accompanied by a rapid decrease of the transparency of the magnetic fluid. The chains fracture in the flow and this results in the restoration of the initial transparency. This phenomenon is used in examining the behaviour of chain aggregates in a flow.

Questions for chapter 7

- *What is the condition that determines the affiliation of the fluid to the class of Newtonian fluids? Give examples of Newtonian fluids.*
- *What is the relationship between the tangential stresses and the speed gradient in a non-Newtonian fluid? Give the examples of non-Newtonian fluids.*
- *List the manifestations of the magnetorheological effect in ferro-suspensions. How are they used in practice?*
- *Explain the Shvedov–Bingham equation.*
- *Describe the types of structural rearrangement which can be observed in a magnetic disperse system? How this influences the magnetic susceptibility?*
- *Describe the chain model of the moderately concentrated disperse systems with inter-connected particles ?*
- *What are the special features of the creep mode in the flow of a ferrosuspension?*

8

Mechanical and magnetic properties of nanodispersed systems

8.1. Equation of the magnetic state of a superparamagnetic

The equation of the magnetic state is the analytical dependence of the magnetisation of a material on the strength of the magnetic field and temperature, i.e. $M(H,T)$. The magnetic particles, responsible for the magnetisation process in the magnetic fluid, are in the superparamagnetic state, i.e., their magnetic moment carries out Brownian motion. The application of the external magnetic field results in rapid saturation of the magnetisation of the magnetic fluid in weak and medium-strength magnetic fields $M \sim 100$ kA/m, because the magnetic moment of the single-domain particle is many times greater than the magnetic moments of the individual atoms and this again stresses the suitability of using the term 'superparamagnetic' for these media.

The energy of interaction of the magnetic moment with the external field and the body of the particle in the case of the 'easy axis' anisotropy is determined by the relationship

$$U = -\mu_0 \mathbf{m}_* \mathbf{H} - K_a V_f (\mathbf{m}_* \cdot \mathbf{n})^2 / m_*^2, \qquad (8.1)$$

where K_a is the anisotropy constant; $\mathbf{n}$ is the unit vector defining the direction of the anisotropy axis.

The mechanism of rotation of the magnetic moment of the particle depends on the ratio of the terms in this expression.

If $\mu_0 m_* H \ll K_a V_f$, the magnetic moment is rigidly connected with the easy magnetisation axis, i.e., it is 'frozen' into the body of the particle. In this case, the mechanism which determines the rotation of the magnetic moment is the rotation of the particle itself. The orientation of the magnetic moment along easy magnetisation axis is established during damping of the Larmor precession of the magnetic moment:

$$\tau_\gamma = (\beta\omega_\gamma)^{-1} \tag{8.2}$$

where $\omega_\gamma = \mu_0 \gamma_a H_a$ is the Larmor precession frequency and its order of magnitude is equal to 10^9 Hz; $\gamma_e = 1.76 \cdot 10^{11}$ C/kg is the hydromagnetic ratio for the electrode; β is the dimensionless damping parameter equal to 10^{-2}; $H_a = 2K_a/\mu_0 M_s$ is the strength of the anisotropy field.

The thermal rotational fluctuations of the moment result in weakening of its bond with the body of the particle. This mechanism is characterised by a dimensionless parameter

$$\sigma_* = K_a V f / k_0 T, \tag{8.3}$$

equal to the ratio of the anisotropy energy to the energy of thermal fluctuations. The anisotropy constant for ferromagnetics varies in a wide range $K_a \approx (10^2-10^6)$ J/m³. At room temperature $k_0 T \approx 10^{-21}$ J and the particle radius $R \approx 4$ nm we obtain $\sigma_* \approx 10^2 \div 10^{-2}$.

Thus, σ_* may have the values both considerably greater and smaller than the unity.

If $\mu_0 m_* H \gg K_a V_f$, the orientation of the magnetic moment of the particle is close to the orientation of the magnetic field. The relaxation time depends on the frequency of the ferromagnetic resonance determined in this case by the strength of the external field.

In the non-rigid dipoles where the ratio $\mu_0 m_* H / k V_f$ may have arbitrary values, and the orientation of the vectors **H** and **n** is also arbitrary but fixed, the magnetic moment is established along the effective field

$$\mathbf{H}_e = \mathbf{H} + \mathbf{H}_a \tag{8.4}$$

where $\mathbf{H}_a = \dfrac{2K_a}{\mu_0 M_s m_*}(m_* \cdot n)n.$

To take into account the thermal fluctuations of the moment it is necessary to clarify the conditions of existence of rigid dipoles. The fluctuation energy should be considerably smaller than the binding energy of the moment with the body of the particle, i.e. $\sigma_* \gg 1$. However, even if this condition is satisfied in the time period τ_N (Néel relaxation time) characteristic of the given ferroparicles, the particle is remagnetised. Therefore, the particle can be regarded as a rigid dipole over the time period satisfying the condition $t_* \ll \tau_N$.

In the quasistatic approximation the magnetisation vector of the magnetic fluid **M** and the vector of the strength of the magnetic field in the medium **H** are assumed to be parallel. The relationship between the modules M and H is defined by the equilibrium magnetisation equation. At low concentrations of the ferrocolloid the ferroparticles dispersed in it may be regarded as non-interacting Brownian particles taking part in the random thermal motion with the energy $k_0 T$. As a result, the set of these particles assuming the rigid bond of the magnetic moment with the body of the particle can be treated as a gas and the process of its magnetisation can be described by the theory of magnetisation of the paramagnetic gas.

Because of the frequent use of the Langevin model of magnetisation in analysis of different processes taking place in the magnetic fluids, we shall describe the derivation of the equation of the magnetic state of the paramagnetic gas (P. Langevin, 1905).

To describe the angular distribution of the magnetic moments of the particles, we use, as shown in Fig. 8.1, the unit vectors constructed from the origin of the coordinates to the surface of the unit sphere. At $H = 0$ the magnetic moments are oriented in all directions and, therefore, the distribution of the ends of the unit vectors on the surface of the unit sphere is uniform. When the field H is applied the distribution is displaced to the directions close to

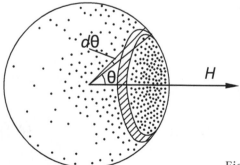

Fig. 8 .1. The Langevin model.

the direction H. When the magnetic moment of the particle forms the angle θ with the field V_p, the probability of the field oriented in this direction is proportional to the Boltzmann factor:

$$\exp\left(-\frac{U}{k_0 T}\right) = \exp\left(\mu_0 \frac{m_* H}{k_0 T} \cos\theta\right). \tag{8.5}$$

The fraction of the magnetic moments, forming with the field angles in the range from θ to $\theta + d\theta$, is proportional to the area of the crosshatched ring in Fig. 8.1 which is equal to $2\pi \cdot \sin\theta \cdot d\theta$. Consequently, we obtain the following equation for the probability of the magnetic moments being oriented in the angle range between θ and $\theta + d\theta$:

$$p(\theta)d\theta = \frac{\exp\left(\mu_0 \dfrac{m_* H}{k_0 T} \cos\theta\right) \sin\theta d\theta}{\displaystyle\int_0^\pi \exp\left(\mu_0 \dfrac{m_* H}{k_0 T} \cos\theta\right) \sin\theta d\theta}. \tag{8.6}$$

If the magnetic moment forms angle θ with the field H, its component along the field is equal to $m_* \cdot \cos\theta$ and, therefore, the magnetisation, determined by all magnetic moments, present in the unit volume, is

$$M = nm_* \overline{\cos\theta} = nm_* \int_0^\pi \cos\theta\, p(\theta)d\theta = nm_* \frac{\displaystyle\int_0^\pi \exp\left(\mu_0 \dfrac{m_* H}{k_0 T} \cos\theta\right) \cos\theta \sin\theta d\theta}{\displaystyle\int_0^\pi \exp\left(\mu_0 \dfrac{m_* H}{k_0 T} \cos\theta\right) \sin\theta d\theta}. \tag{8.7}$$

Carrying out integration in the numerator and the denominator of this formula, and assuming $\mu_0 m_* H / k_0 T \equiv \xi$, gives

$$M = nm_* \left(\frac{e^\xi + e^{-\xi}}{e^\xi - e^{-\xi}} - \frac{1}{\xi}\right) = nm_* \left(\mathrm{cth}\,\xi - \frac{1}{\xi}\right). \tag{8.8}$$

The equation in the brackets is referred to as the Langevin equation and is denoted by $L(\xi)$.

Thus, the theory leads to the magnetisation law described by the Langevin equation:

$$M = nm_*L(\xi),\ L(\xi) = cth\xi - \frac{1}{\xi},\ \xi = \frac{\mu_0 m_* H}{k_0 T}. \qquad (8.9)$$

The thermal fluctuations determine the stochastic rotation of the magnetic moment in relation to the direction of the field. The effect of this mechanism on the orientation of the magnetic moment is determined by the Langevin argument ξ. The conventional boundary between the 'weak' and 'strong' magnetic fields for the magnetic fluid may be the value $H_T = k_0 T/\mu_0 m_*$, derived from the condition $\xi = 1$.

We estimate H_T. For magnetite particles with the characteristic volume $V_f = 5 \cdot 10^{-23}$ m^3 the magnetic moment $m_* = M_s V_f = 2.25 \cdot 10^{-19}$ A/m^2. At the energy $k_0 T = 4.15 \cdot 10^{-21}$ J we obtain $H_T = 1.46 \cdot 10^4$ A/m. Therefore, in investigating the magnetisation of the magnetic fluid the magnetic fields up to ~15 kA/m are conventionally classified as 'weak'.

With increase of the strength of the magnetic field the curve $L(\xi)$ approaches asymptotically the unity which corresponds to the saturation magnetisation of the medium $M_S = nm_*$, i.e., the total orientation of the magnetic moments of all particles along the field. In strong magnetic fields, when $H \gg k_0 T/\mu_0 m_*$, equation (8.9) has the form

$$M = M_S - \frac{3M_S k_0 T}{4\pi\mu_0 M_{S0} H R^3}, \qquad (8.10)$$

where M_{S0} is the saturation magnetisation of the dispersed ferromagnetic; R is the radius of the ferroparticles.

In weak fields in expanding the Langevin equation into a Taylor series, we obtain

$$\lim_{\xi \to 0} L(\xi) = \xi/3. \qquad (8.11)$$

and, consequently, the initial magnetic susceptibility $\chi_0 = \dfrac{M}{H}$ does not depend on the strength of the field

$$\chi_0 = \frac{4\pi\mu_0 M_S M_{S0} R^3}{9k_0 T}. \qquad (8.12)$$

The equations (8.10) and (8.12) show that the superparamagnetism of the magnetic fluid is interesting not only as a specific magnetic phenomenon but also as a non-destructive method of determination of the dimensions and magnetic moment of the magnetic nanoparticles dispersed in the colloid.

The numerical value of the initial magnetic susceptibility of a concentrated magnetic fluid (volume concentration of magnetite ~0.2) at room temperature reaches ~10, which is ~10^4 times higher than the susceptibility of the ordinary fluids.

The value of χ_0 decreases with increasing temperature. When the temperature approaches the Curie point T_k of the magnetic from which the colloid is produced, its spontaneous magnetisation also shows a strong dependence on temperature. Heating the magnetic fluid above T_k can greatly reduce its magnetic susceptibility and this is the basis of the phenomenon of thermomagnetic convection. The layers of the magnetic fluid with $T < T_k$ are characterised by high magnetic susceptibility and are 'pulled' into regions with higher strength of the magnetic field, displacing the layers with $T > T_k$. The intensity of thermomagnetic convection is many times greater than that of gravitational convection.

The susceptibility increases in accordance with the Curie–Weiss law with a decrease of temperature T, but this increase is not without limits and at some temperature T_g the dependence $\chi_0(T)$ reaches a maximum followed by a decreased of χ_0. The numerical value of T_g is not associated with the solidification temperature of the liquid carrier, and depends on the concentration φ of the magnetic substance of the magnetic fluid and the frequency of the measurement field. With decreasing temperature the system of the interacting magnetic dipoles – single-domain colloidal particles forms a random structure of dipoles chains entangled in a complicated manner. For example, at $T > T_g$ the magnetic fluid is a liquid superparamagnetic, and at $T < T_g$ its changes to the disordered gel-like state.

The equation (8.9) shows that in narrow ranges of the variation of temperature, the strength of the magnetic field and the concentration, the equilibrium value of the magnetisation of the compressible magnetic fluid M_e can be represented by a linear dependence:

$$M_e = M_0 + M_n \cdot \delta n + M_T \cdot \delta T + M_H \cdot \delta H, \qquad (8.13)$$

where $M_0, M_n \equiv \left(\dfrac{\partial M}{\partial n}\right)_0$, $M_T \equiv \left(\dfrac{\partial M}{\partial T}\right)_0$, $M_H \equiv \left(\dfrac{\partial M}{\partial H}\right)_0$ relates to the non-perturbed medium.

The equation (8.13) does not contain the term responsible for thermal expansion. The point is that the variations of temperature δT in this case are a consequence of the adiabatic nature of deformation $\partial u/\partial x$. In this case, the medium in the conditions of uniform compression does not undergo thermal expansion (a temperature jump in the medium with the absolutely rigid wall of the vessel).

In addition to M_H we also use the concept 'total' or 'integral' magnetic susceptibility $\chi = M/H$.

8.2. Diffusion of nanoparticles in a fluid matrix

The nanoparticles of the disperse phase are so small that they are suspended in the liquid matrix and carry out Brownian thermal motion. The mean square of the distance over which the particle is removed from the initial point during time t is

$$\langle r^2 \rangle = 6Dt, \qquad (8.14)$$

which shows that the average distance, travelled by the particle during time t is proportional to the square root of this time. The diffusion coefficient D can be calculated from the mobility of the suspended particles b:

$$D = k_0 Tb. \qquad (8.15)$$

For spherical particles (radius R), the mobility is $b = (6\pi\eta R)^{-1}$. Therefore

$$\langle r^2 \rangle = k_0 Tt \, (\pi\eta R)^{-1} \qquad (8.16)$$

If it is assumed that $T = 300$ K, $t = 10^{-6}$ s (the period of the ultrasound wave), $\eta = 1.38 \cdot 10^{-3}$ Pa · s, $R = 5 \cdot 10^{-9}$ m, then $\langle r \rangle \approx 5.5 \cdot 10^{-9}$ m, because $\lambda \approx 1$ mm.

The order of magnitude of time τ during which the particles suspended in the liquid rotates around its axis as a result of Brownian motion over a 'large' angle is:

$$\tau \approx \eta R^3 (k_0 T)^{-1} \approx 4 \cdot 10^{-8} \text{ s}. \qquad (8.17)$$

Thus, the equation of immobility of the disperse magnetic nanoparticles in the liquid carrier in the ultrasound frequency range is fulfilled only for the translational motion.

The nanoparticles with mass m at room temperature move with the thermal velocities $v = \sqrt{2k_0T / m} \approx 1.7\,\text{m/s}$, and the characteristic time during which the particle changes the direction of motion is $t \sim m/3\pi\eta d_0 \sim 10^{-10}$ s. During this time the particle travels over the distance ~ 0.1 nm. Carrying out fast random motion with the 'step' of ~ 0.1 nm, the particle diffuses slowly, moving on average over the distance $(2Dt)^{\frac{1}{2}}$ during time t. The equilibrium distribution of the concentration of the particles is established during the finite time τ_*. Its order of magnitude is determined by the characteristic diffusion time:

$$\tau_* = 6\pi\eta R k_0 T/f^2. \qquad (8.18)$$

where f is a force acting on the particle (in this case $f = mg$); η is the viscosity of the liquid carrier; R is the particle radius.

If it is assumed that $\eta = 1.38 \cdot 10^{-3}$ Pa $\cdot$ s (viscosity of kerosene), $R = 5$ nm, $T = 300$ K, $\rho = 5240$ kg/m^3 (the density of magnetite), then $\tau_* \approx 7 \cdot 10^7$ s ≈ 23 years.

In the field of the gravitational force this system is remains homogeneous as long as necessary.

Taking into account the presence in the real magnetic fluids of the aggregates in the form of the magnetic fluid chains, it should be noted that in the ultrasound and magnetic fields the orientation of the magnetic chains is determined by three factors: the magnetic field, thermal motion and the velocity of the carrying fluid [35].

The body with the homogeneous flow of the ideal, incompressible liquid around it, is subjected to the effect of the moment of the force is equal to:

$$M_r = -\frac{1}{2}\left(\lambda_{\parallel} - \lambda_{\perp}\right)U^2 \sin 2\theta, \qquad (8.19)$$

where $\lambda_{\parallel}$ and $\lambda_{\perp}$ are the components of the tensor of the attached masses of the ellipsoids; U is the flow-around rate; θ is the angle between the direction of the speed **U** and the major axis of the solenoid.

The efficiency of the rotational influence of the fluid flow on the aggregates in comparison with the effect of the thermal Brownian motion and the magnetic field is relatively small. This is indicated by the estimates of the strength of the magnetic field at which the rotational effect is reached, with this effect being characteristic of the ultrasound wave of average power, with comparison of the energies of the rotational effect of the flow with the thermal energy of the particles.

Using the expressions $\lambda_\parallel$ and $\lambda_\perp$ for the ellipsoids of rotation with the major and minor axis ℓ and d, for the order of magnitude we obtain: $M_r \sim (4/3)\pi\rho_1 \ell d^2 U^2$. The magnetic moment of the ellipsoids is determined from the equation

$$m = m_* N_{ag} = M'_S 4\pi \ell d^2/3, \qquad (8.20)$$

where N_{ag} is the number of ferromagnetic particles in the aggregate, m_* is the magnetic moment of a single ferromagnetic particle; M'_S is the magnetisation of ferromagnetic particles.

We estimate the magnitude of the magnetic field at which the equality $\dfrac{M_r}{\mu_0 mH} = 1$ is satisfied. We derive relationship

$$\frac{M_r}{\mu_0 mH} = \frac{(4/3)\pi \ell d^2 U^2 \rho_1}{(4/3)\mu_0 \pi \ell d^2 M'_S H} = \frac{U^2 \rho_1}{\mu_0 M'_S H}. \qquad (8.21)$$

For the ultrasound with a power of 1W/cm² at a frequency of $\nu = 1$ MHz the amplitude of displacement is of the order of $2 \cdot 10^{-8}$ m. This corresponds to the amplitude of the vibrational velocity $U = 0.13$ m/s. Assuming that $M'_S \sim 4.7 \cdot 10^5$ A/m, we obtain that $M_r = \mu_0 mH$ at $H \approx 20$ A/m.

We estimate the volume of the ellipsoidal particle, having the energy in the flow comparable with the energy of the thermal Brownian motion ($U \sim 0.02$m/s, $T \sim 300$ K)

$$V_{ag} = \frac{4}{3}\pi \ell d^2 = \frac{k_0 T}{\rho_1 U^2} \approx 1.3 \cdot 10^{-20} \text{m}^3. \qquad (8.22)$$

Assuming for the order of magnitude of $V_{ag} = N_{ag} \cdot 4\pi R_m^3/3$ (R_m is the radius of a single ferromagnetic particle), we obtain the estimate of the number of particles in the aggregate N_{ag}: $N_{ag} \approx 2 \cdot 10^4$. Consequently, the Brownian motion can be ignored only in the case of the relatively large dimensions of the aggregates.

8.3. Magnetodiffusion and barodiffusion in nano- and microdispersed media

The magnetisation of the real magnetic fluids and ferrosuspensions may be characterised by the rearrangement of the structure in which the microscopic homogeneity of the system is disrupted. These disruptions of homogeneity may be caused by the magnetodiffusion of ferroparticles. The directional magnetodiffusion under the effect of an inhomogeneous magnetic field is referred to as *magnetophoresis*.

We estimate the speed of displacement of the ferroparticles in an inhomogeneous magnetic field in a disperse system with relatively large particles, ~1 μm, for which the following condition is satisfied: $k_0T/\mu_0m_*Gh \ll 1$ [1]. To satisfy this condition, it is necessary to ignore the processes of diffusion of the microparticles.

We assume hypothetically that in a fluid there is a cube with the face h. The equilibrium concentration of the particles in the absence of the field is n_e. In an inhomogeneous magnetic field, each particle is on average affected by the force $\mathbf{F}_1 = \mu_0m_*L\mathbf{G}$. Let it be that G is co-linear with the axis Z and normal to one of the faces of the cube, and its magnitude increases linearly along the length h from G_1 to G_2. In the quasi-stationary mode the particles carry out directional motion along G with a constant speed ϑ, and the value of this speed is determined from the following relationship

$$\vartheta = \mu_0 M_{SO}V_0L(\xi)G / 6\pi\eta_S R. \qquad (8.23)$$

During the time Δt, ΔN_1 particles are supplied through the left face of the cube, and the number of particles passing through in the same time through the right face is ΔN_2. In this case $\Delta N_1 = n_e\vartheta_1h^2\,\Delta t$ and $\Delta N_2 = n_e\vartheta_2h^2\,\Delta t$, i.e., the value of n differs only slightly from n_e within the limits of the investigated cube. Consequently, in the specified cube, the number of the particles decreases by ΔN which, in the first approximation, is uniformly distributed in the volume of the cube. Consequently, the increase of the concentration is

$$\Delta\varphi = -2\mu_0 n_e M_{SO}V_0L(\xi)R^2\Delta t / 9\eta_S h. \qquad (8.24)$$

In the disperse systems there may also be gravitational diffusion (*barodiffusion*). The ratio of the magnetic force, acting on the particle, to the gravitational force β' does not depend on the size of the particles:

$$\beta' = \mu_0 M_{S0} G / (\rho_2 - \rho_1) g. \qquad (8.25)$$

At $\beta' > 1$ the magnetic force is greater than the gravitational force, and at $\beta' < 1$ the gravitational force is greater than the magnetic force.

The character of the magnetic effect greatly differs from the effect of gravitation owing to the fact that the magnetic effect can be used to obtain the variable, as regards intensity and direction, magnetic transfer of the magnetic material in the finite volume, for example, in an acoustic cuvette, which in turn enables us to construct the required geometry of the distribution of concentration and the speed of sound in the volume and control the cross-section and direction of the sound beams.

The redistribution of the concentration of the disperse phase in the volume results in the formation of a gradient of the speed of sound and, consequently, the refraction of the sound beams.

8.4. Aggregative stability of the disperse system of magnetic nanoparticles

The fluid ferromagnetics, synthesised in the middle of the 60s of the 20th-century, represent colloidal solutions of different ferro- and ferrimagnetic substances in ordinary fluids. In the formation of the magnetic fluids it is possible to solve one of the most important problems of colloidal chemistry – the formation of nanoparticles of a solid material and its dispersion in a liquid carrier. At such small dimensions the particles become single-domain. In the absence of the magnetic field and in fields in which the paraprocess is not significant, the single-domain particles can be regarded as magnetised to saturation. Their magnetic moment is $m_* = V M_{S0}$, where V is the volume of the particle. The saturation magnetisation M_{S0} depends on the size of the particles and decreases with a decrease of the particle size; at the particle size M_{S0}, typical of the magnetic colloids, it is equal to ~50% of the appropriate value of the multi-domain material. The decrease of M_{S0} is linked with the shortage of the neighbours in the exchange interaction in the surface layer or chemical changes of the surface layer of the particles.

The particles in the colloidal solution interact by van-der-Waals forces which are short-range and strong only if the particles are close to each other. They are regarded as the surface forces. In addition to the surface forces, there are also forces determined by the presence in the particles of a constant magnetic moment acting between the particles of the magnetic colloids [1]. The energy of the dipole interaction of the pair of the identical ferroparticles can be described by the following equation,

$$U = -2\mu_0^2 \, m_*^4/(3r^6 k_0^T), \tag{8.26}$$

where μ_0 with the magnetic constant; m_* is the magnetic moment of the particle; r is the distance between the particles; k_0 is the Boltzmann constant; T is absolute temperature.

The energy of magnetic interaction decreases with increasing distance at a considerably lower rate than the energy of the van-der-Waals interaction, i.e., the magnetic forces are long-range. When the particles come closer to each other, they start to merge, resulting in the aggregation of the disperse phase so that the colloidal solution loses its stability. The condition of existence of the magnetic fluid as a stable colloidal solution is reduced due the fact that the energy of the magnetostatic interaction of the magnetic dipoles U represents a small fraction of the thermal energy of the particles $k_0 T$.

The aggregate stability of the colloids is achieved by forming, on the surface of the particles, shielding shells preventing bonding of the particles into aggregates.

The saturation magnetisation and the stability of the magnetic fluid are strongly influenced by the special features of the processes of synthesis of the magnetic fluid: the rate of supply of the solution of ferrous salts to the alkali and the mixing intensity of the reaction mixture, the selection of the settling agent, temperature conditions. In addition to this, the properties of the final product are also influenced by the 'internal' factors of the technological process: the type and concentration of the magnetic phase, the size of the ferroparticles, the method of stabilisation, the type and concentration of the stabiliser, the composition of the liquid carrier, the presence of different additions. The number of the factors and also special features of the synthesis technology, influencing the properties of the actual magnetic fluids, is so large that they are very difficult to control and produce fluids with the required properties.

The interaction of the particles in the magnetic colloids under specific conditions results in the formation of the structure of ferroparticles (floccules, granules, chains, clusters, spatial network, the pre-shaped aggregates) which has a strong effect on the magnetic properties of the magnetic fluid. There are two mechanisms influencing the coalescence of magnetic colloids – molecular attraction between the suspended particles and the dipole–dipole interaction specific for the ferroparticles. To characterise the aggregative stability of the system, the constant of 'pairing' of the particles with the diameter d was introduced:

$$\Pi = \mu_0 m_*^2/(d^2 k_0 T). \tag{8.27}$$

In the single-domain particles Π is proportional to the volume of the particle. At $\Pi < 1$ the controlling role is played by the van-der-Waals forces. The contribution of the magnetic interaction to the general balance of the interparticle forces increases with increase of the particle size. At $\Pi \geq 1$ the magnetic attraction of the particles leads to the formation of spatial structures – *chains and clusters* – as a result of the appearance of the minimum of the total energy of interaction of the particles at large distances between them.

The magnitude of the magnetic susceptibility of the magnetic fluid increases with increase of the size of the magnetic particles and their volume concentration φ in the colloids. The particle size $d \sim 10$ nm is optimum because this is the largest size at which the particle does not yet aggregate because of the dipole–dipole interaction at room temperature (bonding of the particles is prevented by the thermal motion of the particles). The maximum concentration of the magnetic substance φ_m in the colloid depends on the ratio δ/d and the size distribution of the particles (δ is the thickness of the stabilising shell). If all the particles were identical spheres with a diameter d, then at dense packing of the particles in the hexagonal or face-centred cubic lattice the value $\varphi_m = \left(\pi/3\sqrt{2}\right)\left[d_0/\left(d_0 + 2\delta\right)\right]^3$ would be ≈ 0.27 at $d_0 = 10$ nm and $\delta = 2$ nm. Usually, the magnetic fluid contains particles of different sizes and they can be packed with higher density. The concentration of the magnetic phase in the magnetic fluid may reach 0.3 but, in most cases, in the magnetic colloids $\varphi_m \approx 0.1$–0.2, and the amount by which the viscosity is greater than the viscosity of the liquid carrier is in the range of several percent to several orders of magnitude.

The repulsion force, referred to as the steric force, forms between the particles coated with a layer of long chain molecules.

Steric repulsion forms as a result of the distortion of the long molecules and the increase of their local concentration in the zone of intersection of the solvate layers. In long-term contact of the particles, the excess of the flexible links of the surfactant molecules may be redistributed over a large volume or over the entire adsorption layer at a sufficiently high surface mobility of the adsorbed molecules. At the same time, the repulsion forces depend on the contact time of the particles indicating that these forces are not purely potential. As regards the nature of the resistance of the particles to convergence, the adsorption layers should be regarded as elastoviscous shells with the elasticity modulus dependent on the repulsion potential, and the relaxation time of the stress, determined by the rate of establishment of the equilibrium distribution of the molecules in the adsorption layer.

In the field, the minimum energy of interaction of the particles corresponds to the angle $\hat{m_* r} = 0$ and, consequently, the particles should be distributed in chains along the field. In a powerful magnetic fields U may increase by orders of magnitude in comparison with the interaction energy outside the field. Therefore, in the colloids in which there is no spontaneous agglomeration of the particles, the external magnetic field may cause reversed agglomeration. This is confirmed by, for example, the effect of the field on the optical transparency of the colloids of magnetite or the anisotropy of scattering of the light which change when a field is applied and restore their initial values when the field is switched off.

In the presence of the aggregative stability of the system, the particles of the disperse phase are, because of their small size, sustained by the thermal Brownian motion in the volume of the liquid carrier.

The magnetic fluids show high stability also in the magnetic field with a strong heterogeneity. In this case, the ferroparticles are subjected to the effect of the on the to forced

$$f = \mu_0 m_* G, \tag{8.28}$$

where G is the gradient of the strength of the magnetic field, m_* is the magnetic moment of the particle.

Assuming that $m_* \approx 10^{-19}$ A $\cdot$ m^2, $G = 10^6$ A/m^2, we obtain $\tau_* \approx 6 \cdot 10^6$ s ≈ 60 days, which is the estimate of the delamination time of the magnetic fluid.

The magnetic fluids are practically opaque fluids. The experiments with translucence can be carried out either if the thickness of the layer is small (~10 μm) or in the case of a low concentration ($\leq 10^{-2}$) with a thickness of the layer being ~1 mm.

In the electrical or magnetic fields, the magnetic fluids are similar to uniaxial crystals. They show the anisotropy of thermal and electrical conductivity, viscosity, and also the anisotropy of the optical properties: birefringence, dichroism and the anisotropy of scattering. These effects are associated with the orientation along the external magnetic field **H** or electrical field **E** of non-spherical colloidal particles and also with their alignment into high-density chains directed along the field. The characteristic values of the strength of the electrical and magnetic fields at which the orientation effect becomes strong can be estimated, equating the electrostatic or magnetostatic energies for the particles of the average volume $V_f = 5 \cdot 10^{-23}$ m^3 to the energy of its thermal motion: $\mu_0 m_* H \approx k_0 T$ or $\varepsilon_0 V_f E^2 \approx k_0 T$. From this we can obtain $H \approx k_0 T / \mu_0 m_* \approx 1.46 \cdot 10^4$ A/m and $E_0 \approx (k_0 T / V_f)^{1/2} \approx 3 \cdot 10^6$ V/m.

The values of the electro- and magneto-optical effects in the magnetic fluids are six orders of magnitude higher than the identical values in the ordinary fluids because the volume of the colloidal the particles is ~10^6 times greater and the volume of the molecules. In the crossed electrical and magnetic fields, the magnetic fluids are similar to biaxial crystals in which the optical anisotropy can be changed both in the magnitude and direction. At a specific ratio between **H** and **E**, directed normal to each other, the effect of compensation of optical anisotropy is observed. This takes place at $H/E \approx 5 \cdot 10^{-3}$ ohm^{-1}.

Questions for chapter 8

* *Name the physical quantities linking the equation of the magnetic state of the superparamagnetic?*
* *What is the difference between the Brownian and Néel mechanisms of magnetisation of the particles? What are the characteristic dimensions of the particles for each mechanism?*
* *Which model is used for deriving the equation of the magnetic state the magnetic fluid?*
* *Write the Langevin function. How is this function linked with the law of magnetisation of the magnetic fluid? What is the parameter of the Langevin function?*
* *Name the criterion determining the magnetic field as weak? What is approximately its value equal to?*
* *What is the limit of the Langevin function equal to, if the argument*

tends to 0? If the argument tends to infinity?
- *Why is the magnetic susceptibility a constant quantity in weak magnetic fields?*
- *Describe the phenomenon of thermomagnetic convection?*
- *Explain the principle of magnetodiffusion and barodiffusion in magnetic disperse systems*
- *What is the aggregative stability of the dispersed system of magnetic nanoparticles?*
- *What is the steric repulsion of the surfactant molecules?*

9

Additive model of the elasticity of nano- and microdispersed systems

9.1. The additive model of the elasticity of nano- and microdispersed systems taking interphase heat exchange into account

To obtain the functional dependence of the rate of propagation of the sound waves in a disperse medium with solid nanoparticles of the disperse phase (for definition – in a magnetic fluid) on the volume concentration of the particles φ, we can use the 'additive' model of the formation of the elasticity of microheterogeneous systems proposed by V.M. Polunin [36].

The calculation equation will be derived assuming that there is no heat exchange between the components of the hypothetical disperse system. It will also be assumed that the densities of the stabiliser and the liquid carrier are approximately equal to each other and, therefore, we can use the relationship:

$$\rho = \rho_1(1- \varphi) + \rho_2 \varphi \qquad (9.1)$$

where ρ is the density of the fluid; φ is the concentration of the solid phase.

The volume of the magnetic fluid consists of the volumes of the liquid carrier V_1, the solid phase V_2 and the stabiliser V_α. To obtain the highest stability of the system the concentration of the stabiliser should be optimum. Let it be that the value $\alpha \equiv V_\alpha/V_2$ satisfies these requirements and remains constant for some class of the magnetic fluids.

In the case of the quasistatic increase of the internal pressure by Δp, the volume of the system changes by ΔV, and $\Delta V = -\beta_T V \cdot \Delta p$, where β_T is the isothermal compressibility of the system.

The increment of the volume of the system should be equal to the sum of the increments of the volumes of each component:

$$\Delta V = \Delta V_1 + \Delta V_2 + \Delta V_\alpha \qquad (9.2)$$

where ΔV_1, ΔV_2, ΔV are the increments of the volume of the disperse medium, the solid phase and the stabiliser, respectively.

Consequently, we find

$$\beta_T = (1 - \varphi - \alpha\varphi)\, \beta_{T1} + \alpha\varphi\beta_T\alpha + \varphi\beta_{T2}, \qquad (9.3)$$

where β_{T1}, $\beta_{T\alpha}$, β_{T2} as the isothermal compressibilities of the disperse medium, the stabiliser and the solid phase; $\varphi \equiv V_2/V$ is the bulk concentration of the solid phase.

Taking into account the relatively small compressibility of the solid bodies, it is assumed that $\beta_{T2} = 0$. Various assumptions have been made as regards the parameters $\beta_{T\alpha}$ and β_{T1}: $\beta_{T\alpha} \ll \beta_{T1}$ and $\beta_{T\alpha} \approx \beta_{T1}$. The first case is implemented in the presence of a relatively strong bond of the molecules of the stabiliser with the surface of the particles, the second case – in the absence of such a bond. It will be more accurate to assume that $\beta_{T\alpha} = \gamma\beta_{T1}$, where γ is the proportionality coefficient, showing the relationship between the compressibility of the stabiliser and the compressibility of the liquid carrier in the disperse system. Consequently, the equation (9.3) can be determined in the following form

$$\beta_T = [1 - \varphi - (1 - \gamma')\,\alpha\varphi]\, \beta_{T1} \qquad (9.4)$$

In high-speed processes, the isothermal compressibilities β_{T1}, β_{T2} and $\beta_{T\alpha}$ change to adiabatic compressibilities β_{S1}, β_{S2} and $\beta_{S\alpha}$, which have the real value in the absence of the dissipative processes.

If the mutual effect of the components is ignored, the adiabatic compressibility of the system is the sum of the specific adiabatic compressibilities:

$$\beta_S = [1 - \varphi - (1 - \gamma')\,\alpha\varphi]\, \beta_{S1} \qquad (9.5)$$

If there is a satisfactory agreement with the experimental results, the resultant value the parameter γ' may provide information on the relative compressibility of the components of the actual magnetic fluid.

The concentration of the solid phase in the magnetic fluid is calculated from the equation:

$$\varphi = (\rho - \rho_1)/(\rho_2 - \rho_1). \tag{9.6}$$

The Soviet theoretical physicist M.A. Isaakovich developed in 1948 the theory [37, 38] in which it was shown that the passage of a sound wave in a disperse system is accompanied by periodic compression and tension of the components of the system, and on the 'macroscopic' scale this process, as in the majority of homogeneous fluids, takes place by the adiabatic mechanism. However, the change of the temperature of the components system, determined by changing sound pressure, differs. As a result of the low value and the relatively high heat conductivity of the ferroparticles, their temperature will manage to become equal to the temperature of the liquid carrier and, therefore, the process will be 'microscopically' isothermal. The critical frequency below which the frequency range corresponding to this process is found, is determined from the equation:

$$v_{cr} = \frac{\chi_2}{\pi\rho C_{p_2} R^2}, \tag{9.7}$$

where χ_2 and C_{p2} are the heat conductivity and specific heat at a constant pressure of solid particles; R is their radius; ρ is the density of the magnetic fluid.

For low-concentration magnetic fluids $v_{cr} \approx 10^{11}$ Hz. In conventional emulsions the process of propagation of sound can be 'microscopically' both adiabatic and isothermal. In the additive model, the adiabatic compressibility with the interphase heat exchange taken into account can be represented by a linear function of φ:

$$\beta_{ST} = (1-\varphi)\beta_{S1} + \rho_2 C_{p2} T\left(\frac{q_2}{\rho_2 C_{p2}} - \frac{q_1}{\rho_1 C_{p1}}\right)^2 \varphi. \tag{9.8}$$

Assuming the values for $\rho = 1230$ kg/m³, $\rho_1 = 800$ kg/m³, $c =$

1200 m/s, C_{p1} = 2 kJ/(kg K), C_{p2} = 0.655 kJ/(kg K), q_1 = 9.5 · 10^{-4} K^{-1}, q_2 = 11.4 · 10^{-6} K^{-1}, ρ_2 = 5.21·10^3 kg/m^3, φ = 0.1, T = 300 K, and also taking into account the expression $c = \sqrt{1/\rho\beta_{S1}}$ we obtain the numerical value: β_{ST} = (1.125 + 0.035) 10^{-9} Pa^{-1} – the adiabatic compressibility, characteristic of the kerosene-based magnetite magnetic fluid.

9.2. Compressibility of a magnetite–water magnetic fluid

In most fluids the adiabatic compressibility increases monotonically with temperature. However, water is an exception: its compressibility decreases with increasing temperature (to approximately 60°C). The curve of the temperature dependence of the adiabatic compressibility of water has a characteristic minimum. Experiments show that a magnetic fluid in which water is the liquid carrier and the magnetic phase is nanodispersed magnetite, also has a minimum on the temperature dependence of adiabatic compressibility. With increasing concentration of the magnetic fluid the minimum is displaced to lower temperatures [23]. Figure 9.1 shows a network of $\beta_S(T)$ experimental curves for magnetic fluids of different concentrations. The individual numbers correspond to the following concentrations of the samples: 1 – φ = 0%; 2 – φ = 1.66%; 3 – φ = 2.71%; 4 – φ = 6.12%.

In the adiabatic model the adiabatic compressibility of the 'liquid carrier, the magnetic nanoparticles, stabiliser' system taking interphase heat exchange into account can be represented by a linear function of φ:

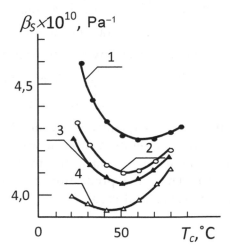

Fig. 9.1. $\beta_S(T)$ curves for a water-based magnetic fluid.

$$\beta_{ST} = \left(1 - \varphi - \alpha\varphi\right)\beta_{S1} + \alpha\varphi\beta_{S\alpha} + \rho_2 C_{p2} T \left(\frac{q_2}{\rho_2 C_{p2}} - \frac{q_1}{\rho_1 C_{p1}}\right)^2 \varphi, \quad (9.9)$$

where α is specific concentration, $\beta_{S\alpha}$ is the adiabatic compressibility of the stabilising substance.

The sum of the first two terms is the adiabatic compressibility of the system without considering the heat exchange between the phases, the last term is the addition due to internal heat exchange.

We obtain information on the displacement of the temperature minimum of the dependence of the adiabatic compressibility of the water-based magnetic fluid with a concentration of the solid phase using the additive model. In some studies this problem is not completely solved.

The adiabatic compressibility of the water-based magnetic fluid can be written in the form:

$$\beta_{ST} = \left(1 - \varphi - \alpha\varphi\right)\beta_{S1} + \alpha\varphi\beta_{S\alpha}. \quad (9.10)$$

Here we have omitted the term associated with internal heat exchange as the value of the volume expansion coefficient of the liquid carrier is small. The minimum compressibility condition is:

$$\frac{\partial\beta_{ST}}{\partial T} = \left(1 - \varphi - \alpha\varphi\right)\frac{\partial\beta_{S1}}{\partial T} + \alpha\varphi\frac{\partial\beta_{S\alpha}}{\partial T} = 0. \quad (9.11)$$

For water $\frac{\partial\beta_S}{\partial T}\langle 0$, and for all other fluids $\frac{\partial\beta_S}{\partial T}\rangle 0$. If it is assumed, as postulated in the additive model, that the surfactant adsorbed in the shells is in the liquid state, characterised by the 'normal' temperature dependence of the compressibility, then $\frac{\partial\beta_{ST}}{\partial T} = 0$ already at some value $\frac{\partial\beta_{S1}}{\partial T}\langle 0$ (i.e., in the temperature range in which β_1 continues to decrease with increasing T). Consequently, the value β_{ST} passes through the minimum at a lower temperature than in water, free from the disperse phase. This conclusion is confirmed by the experimental data shown in Fig. 9.1. Thus, the prediction of the additive model is confirmed by the experimental results.

Questions for chapter 9

- *What are the assumptions of the additive model of elasticity of the magnetic disperse system?*
- *The contribution of which components should be taken into account in calculating the compressibility of the magnetic fluid within the framework of the additive model of elasticity?*
- *Is the process of propagation of the sound wave in the magnetic fluid adiabatic or isothermal? Please explain the answer.*
- *Why are the water-based magnetic fluids regarded as a special case for the additive model of elasticity?*
- *State whether the additive model of elasticity of the magnetic disperse system has been confirmed by experiments?*

'Suspending' a magnetic fluid in a pipe to investigate its rheological properties

10.1. The column of the magnetic fluid in a pipe as an inertia–viscous element of the oscillatory system

We generalise the method of calculating the elasticity coefficient, described in section 5.5, for the case of an oscillatory system with an inertia element in the form of a column of a magnetic fluid (the calculation diagram is shown in 5.8).

The inertia element of the oscillatory system is a column of the magnetic fluid in a pipe sustained above a closed air cavity as a result of the stabilisation of the fluid–gas interface by a heterogeneous magnetic field. The open upper surface of the fluid is outside the limits of the magnetic fields.

The coefficient of ponderomotive elasticity k_p is determined using the results obtained in a model with concentrated parameters. In the conditions of this problem, equation (5.88) is transformed to the following form

$$k_p = \mu_0 S \left[(1+\chi) M_z \frac{\partial H_z}{\partial Z} \right]_{Z_f},$$

(10.1)

where Z_f is the coordinate of the lower base of the magnetic fluid–column.

Taking the elasticity coefficient of the gas cavity k_g into account:

$$k = \rho_g c^2 \frac{S^2}{V_0} + \mu_0 S \left[(1+\chi) M_z \frac{\partial H_z}{\partial Z} \right]_{Z_f}.$$ (10.2)

The frequency of free non-damping oscillations of the oscillatory system with the elasticity of this type is expressed by the equation:

$$\nu_T = \frac{1}{2\pi\sqrt{h}} \sqrt{\frac{\rho_g c^2 S}{\rho V_0} + \frac{\mu_0 M_z G_z}{\rho}(1+\chi)},$$ (10.3)

where h is the height of the magnetic fluid–column.

If the closed gas cavity is part of the pipe, the last equation is transformed to the form

$$\nu_T = \frac{1}{2\pi\sqrt{h}} \sqrt{\frac{\rho_g c^2}{\rho h_g} + \frac{\mu_0 M_z G_z}{\rho}(1+\chi)},$$ (10.4)

where h_g is the height of the air cavity.

The hydrostatic pressure remains constant and does not take 'direct' part in the oscillatory process, but it has an 'indirect' effect on the values of the parameters k_p and k_g as a result of the static displacement of the lower base of the magnetic fluid–column. In addition, when a certain height of the magnetic fluid–column is reached, the air bubbles, overcoming the levitation forces, penetrate through the 'magnetic barrier' and, consequently, the insulation of the gas cavity is disrupted.

Attention will be given to the physical mechanisms of energy dissipation in the column of the magnetic fluid carrying out reciprocating movements in the glass pipe containing the fluid in the absence of the magnetic field.

In the shear flow the solid particle is influenced by the moment of the forces causing the particle to rotate thus increasing the viscosity of the fluid. In the suspension, the presence of the spherical particles results in the increase of viscosity in accordance with the Einstein equation:

$$\eta = \eta_0 \left(1 + \frac{5}{2}\varphi \right), \quad \varphi = \frac{4\pi R^3}{3}n,$$ (10.5)

where φ is the ratio of the total volume of all spheres to the total volume of the suspension (this ratio is small), n is the number of particles in the unit volume.

In the presence of internal friction in the fluid, the flow through the pipe at lower speeds takes place in the laminar manner in the form of cylindrical layers moving with different speeds depending on the distance to the wall. The layer, bordering with the wall, remains stationary, and the axial layer moves at the maximum speed. As a result of friction between the layers, moving at different speeds during the reciprocating flow of the fluid through the pipe, additional losses appear.

In the reciprocating flow of the fluid through the pipe the dissipation of elastic energy as a result of the viscosity of the fluid is adequately interpreted on the basis of the concept of the shear wave, introduced by Stokes. If an infinite plane, placed in the fluid, carries out harmonic oscillations in the direction parallel to the plane, then the quasi-wave process described by the following function forms in the vicinity of the flat surface:

$$\dot{U} = \dot{U}_0 e^{\alpha(z-h)} \cdot \cos\left[\omega t + \alpha(z-h)\right], \qquad (10.6)$$

where h is the distance from the surface, counted along the Z axis, perpendicular to the plane; $\dot{U}$ is the speed of displacement of the fluid layer at a distance h from the surface.

The speed of propagation c, the damping coefficient α and the length of the shear wave λ are expressed by the respective equations [39]:

$$c = \sqrt{\frac{2\omega\eta}{\rho}}, \ \alpha = \sqrt{\frac{\omega\rho}{2\eta}}, \ \lambda = 2\pi\sqrt{\frac{2\eta}{\rho\omega}}. \qquad (10.7)$$

The direction of oscillations of the particles in the investigated wave is perpendicular to the direction of propagation. At a distance of $\lambda/(2\pi)$ the amplitude decreases e times, i.e., 'the depth of penetration' of the viscous wave is $\sigma'' = \lambda/(2\pi)$.

We consider the results of the theory of acoustic impedance affecting the sound wave, propagating in a viscous fluid filling the pipe. If the length of the circumference encircling the side surface of the liquid column is smaller than the double length of the viscous wave, i.e., $\pi d < 2\lambda'$, where $\lambda' = 2\sqrt{\pi\eta/\nu\rho}$ or $d < 4\sqrt{\eta/\pi\nu\rho}$, we obtain the following approximate equation for the impedance of the pipe [40]:

$$R' \approx 8\pi\eta b + i\frac{2}{3}\pi^2\rho b d^2 v. \tag{10.8}$$

The expression $r' = 8\pi\eta b$ corresponds to Poiseuille's law for the coefficient of resistance in the laminar flow of the viscous fluid through a narrow pipe. The Poiseuille flow of the fluid is characterised by the parabolic distribution of the speed of the particles in the cross-section of the pipe. In the present case, the speed of displacement of the boundary of the column $\dot{U}$ is the average speed of the particles through the cross-section of the pipe equal to 50% of its maximum value. For narrow pipes the active resistance in equality (10.8) is higher than reactive resistance, and the total resistance is independent of frequency. In this model, the coefficient of damping of the oscillations β' is calculated from the equation

$$\beta' = \frac{16\cdot\eta}{\rho d^2}. \tag{10.9}$$

With increase of d or v at the given η and ρ the approximate equation (10.8) ceases to be valid.

The apparent part of (10.8) is the inertia component $\left(\frac{4}{3}m\omega\right)$. Thus, this model predicts the presence of the attached mass, equal to $m/3$, as a result of the shear viscosity.

At high values of d and v, when $\pi d/2\lambda' > 10$, another approximate equation is obtained for the complex impedance of the pipe:

$$R'' \approx db\sqrt{\pi^3\rho\eta v} + i\frac{\pi^2 d^2 b\rho v}{2}\left(1+\frac{2}{d}\sqrt{\frac{\eta}{\pi v\rho}}\right). \tag{10.10}$$

The active resistance

$$r'' \equiv db\sqrt{\pi^3\rho\eta v}. \tag{10.11}$$

increases with increasing v. The equation (10.8) for the coefficient of resistance was obtained by Helmholtz for the first time.

The second term in the brackets of equation (10.10) is small in comparison with 1 and, therefore, the apparent part can be used in the form $m2\pi v$, which shows that the Helmholtz model does not contain the attached mass. In this case, the damping coefficient is determined from the following expression:

$$\beta'' = \frac{2}{d}\sqrt{\frac{\pi\eta\nu}{\rho}}.$$ (10.12)

The Helmholtz model predicts the increase of the damping factor with frequency.

It should be noted that the equation (10.12) can be derived by a simple procedure on the basis of the expression for energy dissipation in unit time in the unit area of the oscillating plane [12]:

$$\Delta Q = -\frac{U_0^2}{2}\sqrt{\frac{\omega\rho\eta}{2}}.$$ (10.13)

where U_0 is the amplitude of the oscillatory speed.

This equation holds for high frequencies ('short' viscous waves). The value of energy dissipation in a single period on the area of part of the pipe filled with the fluid is:

$$\Delta Q_d = -\frac{U_0^2}{2}\sqrt{\frac{\omega\rho\eta}{2}}\cdot\frac{\pi\cdot d}{\nu}.$$ (10.14)

The logarithmic damping decrement δ:

$$\delta = -0.5\Delta Q_d / \Delta Q_0,$$ (10.15)

where ΔQ_0 is the total mechanical energy of the oscillatory system, i.e., $\Delta Q_0 = \frac{mU_0^2}{2}$. Then $\beta'' = \delta\cdot\nu = \frac{2}{d}\cdot\sqrt{\frac{\pi\eta\nu}{\rho}}$.

Because of the small depth of penetration of the viscous wave σ'' the oscillatory movement of the fluid column is of the 'piston' nature. The flow of the fluid is concentrated in a thin near-wall layer and this increases the effective viscosity of the magnetic fluid with the quasi-spherical aggregates, comparable with σ''.

To prove these considerations, simple calculations will be carried out. The lower boundary of the frequency range in the experimental conditions can be represented by 20 Hz. Consequently, the length of the viscous wave is

$$\lambda = 2\sqrt{\frac{\pi\eta}{\rho\nu}} = 2\sqrt{\frac{\pi\cdot 8.1\cdot 10^{-3}}{1.5\cdot 10^3\cdot 20}} \cong 1.8\cdot 10^{-3} = 1.8\ \text{mm},$$

and the 'depth of penetration' $\sigma'' \cong 0.3$ mm. If $= 80$ Hz, then $\sigma'' \cong 0.15$ mm.

For a suspension with ellipsoidal particles the correction coefficient at φ in equation (10.5) has different numerical values. For example, the numerical values of the coefficient L' in the equation

$$\eta = \eta_0 \left(1 + L'\varphi\right), \quad \varphi = \frac{4\pi ab^2}{3}n, \qquad (10.16)$$

for several values of $a/b = S$ (a and $b = c$ are the half-axes of the ellipsoids) increase on both sides from the value $S = 1$, corresponding to the spherical particles. This is reflected in table 10.1:

Table 10.1

S	0.1	0.2	0.5	1.0	2	5	10
L'	8.04	4.71	2.85	2.5	2.91	5.81	13.6

For spheres $L' = 2.5$.

If the column of the magnetic fluid is sustained above an isolated air cavity, for example, by a heterogeneous magnetic field, then the adiabatic process of compression and tension of the gas during the periodic displacement of the magnetic fluid column from the equilibrium position results in heat exchange between the gas cavity ,on the one hand, and the walls of the pipe and the open surface of the magnetic fluid on the other hand. As a result of the low heat conductivity of the gas medium, the heat exchange firstly takes place in a relatively narrow boundary region and, secondly, is delayed in relation to the oscillations of the magnetic fluid column. This phase shift results in some additional damping of the oscillations.

The damping factor increases as a result of the energy losses due to the emission of elastic waves by the oscillatory system. The presented results of measurements of the damping factor of the oscillations of the magnetic fluid–column in the pipe are approximated by the dependence of the type $\beta_{ex} \sim v^n$, and $n \approx 0.6$, and their absolute values are higher than those calculated from equation (10.12).

In most cases, the magnetic moment of the particle is rigidly connected with the particle itself (exceptions are represented by only very fine particles with the so-called Néel magnetisation mechanism). Therefore, when the magnetic fields determines the orientation of the magnetic moment of the particle, this complicates its free rotation. This in turn results in local gradients of the speed of the fluid–base in the vicinity of particles and increases the effective viscosity of the magnetic fluid ('magnetoviscous' effect). For the magnetic fluids with spherical particles, the additional internal friction in strong fields is determined by the relationship:

$$\Delta\eta_H = 1.5\varphi_g\eta_0 \cdot \sin^2\beta, \tag{10.17}$$

where $\Delta\eta_H$ is the variation of the viscosity coefficient of the magnetic fluid in the magnetic field; η_0 is the viscosity coefficient at $H = 0$; φ_g is the hydrodynamic concentration of the magnetic fluid; β is the angle between **H** and the angular speed of the magnetic particle.

For a diluted magnetic fluid with a single-particle disperse phase in the flow in a flat capillary the increment of viscosity as a result of the magnetoviscous effect is calculated from the equation [1, 41]:

$$\Delta\eta = \frac{3}{2}\varphi\eta\frac{\xi L^2}{(\xi-L)}, \tag{10.18}$$

where $L = \text{cth }\xi - 1/\xi$ is the Langevin function; $\xi = \dfrac{\mu_0 m_* H}{k_0 T}$.

In relatively strong magnetic fields on reaching the magnetic saturation $\xi \gg 1$, $L \to 1$:

$$\Delta\eta = \frac{3}{2}\varphi\eta, \tag{10.19}$$

in the flow of the colloid in a circular capillary in a field normal to the capillary axis:

$$\Delta\eta = \frac{3}{4}\varphi\eta. \tag{10.20}$$

10.2. Rheology of the magnetic fluid with anisotropic properties

V.A. Naletova and Yu.M. Shkel' (1987) [42] described the experimentally observed 'anomalous' increase of the pressure gradient in the flow of a magnetic fluid in a pipe in a field normal to the flow by the model of the magnetic fluid generalised for the case of a medium with anisotropic properties. The anisotropy of the magnetic fluid is determined by the presence in the disperse phase, in addition to the individual particles, of ellipsoidal aggregates fully oriented by a strong magnetic field. It is assumed that the fluid contains both spherical and elongated particles of aggregates, and the aggregates have the same size, and that there is no interaction amongst the particles of the disperse system. It is also assumed that without the field ($H = 0$) the viscosity factor η is calculated by the Einstein equation:

$$\eta_{H=0} = \eta_0(1+2.5\varphi), \qquad (10.21)$$

The relationship of the pressure gradient between the points z_1 and z_2 along the fluid column a magnetic field oriented in the longitudinal $\Delta p'_{1,2\|}$ and transverse $\Delta p'_{1,2\perp}$ directions in relation to the pipe with the average flow speed v_m and the radius of the pipe R is described by a system of equations

$$
\begin{aligned}
&\Delta p'_{1,2\perp}=\alpha\eta_0(1+\varphi L_\perp(S)), \ \Delta p'_{1,2\|} = \alpha\eta_0(1+\varphi L_\|(S)),\\
&\Delta p'_{1,2}|_{H=0} = \alpha\eta_0(1+2.5\varphi), \ \alpha = 8(z_2-z_1)v_m/R^2,\\
&L_\perp=A(S)+0.5S_2(S)+0.5(1+2\lambda)S_1(S),\\
&L_\|=A(S)+S_2(S)+(1-2\lambda)S_1(S), \ A(S) = 8(S^2-1)/(4S^4-10S^2+3SL),\\
&S_2(S)=4(S2-1)/((S^2+1)(4+2S^2-3SL))-A(S)+\lambda^2S_1(S),\\
&S_1(S) = 2(S2-1)(S^2+1)/S(L(2S^2-1)-2S),\\
&L = \ln|(S+(S^2-1)^{1/2})/(S-(S^2-1)^{1/2}|/(S^2-1)^{1/2},\\
&\lambda = (S^2-1)/(S^2+1),
\end{aligned}
$$
$$(10.22)$$

where S is the geometrical parameter of the nanoaggregate having the form of the ratio of the major half-axis of the ellipsoids to the minor half-axis.

Assuming that the aggregates are sufficiently prolate, since $L_\|\sim2$, the expressions for the pressure gradient can be written in the form:

$$
\begin{aligned}
&\Delta p'_{1,2\perp} = \alpha\eta_0(1+13(\varphi-\varphi_1)/4+\varphi_1 L_\perp(S)),\\
&\Delta p'_{1,2\|} = \alpha\eta_0(1+4(\varphi-\varphi_1)+2\varphi_1),\\
&\Delta p'_{1,2}|_{H=0} = \alpha\eta_0(1+2.5\varphi),
\end{aligned}
$$
$$(10.23)$$

where φ_1 is the volume fraction of the ellipsoidal aggregates.

The system of the three equations (10.23) at the known values of the parameters determined by the equations $\Delta p'_{1,2\|}$, $\Delta p'_{1,2\perp}$ and $\Delta p'_{1,2|H=0}$ can be reduced to two equations with two unknown parameters φ_1 and S. It is shown that the pressure gradient $\Delta p'_{1,2\|}$ in the magnetic field parallel to the fluid flow should be slightly smaller than without the field. In the perpendicular field the prolate aggregates greatly

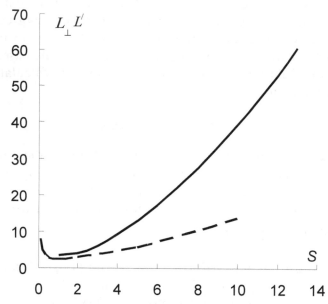

Fig. 10.1. Dependences: $L_\perp(S)$ – solid line; $L'(S)$ – dashed line.

increase the pressure gradient, whereas the spherical particles have no significant effect.

Figure 10.1 shows the theoretical dependence $L_\perp(S)$ and $L'(S)$: $L_\perp(S)$ – solid line, $L'(S)$ – dashed line. The latter was constructed for the 'conventional' (non-magnetic) suspension with ellipsoidal particles according to the data in Table 10.1. The graphic dependence $L_\perp(S)$ was constructed by mathematical modelling in MS Excel starting with the values $S = 2$. This dependence is a monotonically increasing function. Comparison of the curves shows that when reaching the magnetic saturation in the magnetic field transverse to the pipe with the magnetic fluid, the viscosity of the system increases with the increase of the length of the chain far more extensively than in its absence.

Taking into account the proportionality between the effective viscosity η and the pressure gradient $\Delta p'_{1,2\perp}$, the first equation of the system (10.23) can be returning the following form

$$\eta = \eta_0 \left(1 + \varphi L_\perp\right). \qquad (10.24)$$

From the equation of the line of the trend we can find the value $L_\perp(S)$ at $S \rightarrow 1$. In this case, we obtain $L_\perp(S)_{S \rightarrow 1} = 3.44$, which is greater than the appropriate coefficient in the Einstein equation by 0.94 as a result of the magnetic viscosity effect.

The previously examined physical mechanisms of the dissipation of elastic energy in the oscillatory system with the magnetic fluid inertia element can be divided to two groups one of which includes the mechanisms which depend on the strength of the magnetic field (magnetorheological effect) and the other group – the mechanisms independent of the field.

The dissipation of elastic energy in the oscillatory system with the inertia element in the form of the magnetic fluid column (as already mentioned previously) is caused mostly by the simultaneous effect of four physical mechanisms:

1. The energy losses in the reciprocal flow of the fluid through the pipe because in the shear flow on the solid magnetic nanoparticles or ellipsoidal nanoaggregates the nanoparticles are subjected to the effect of the moment of forces; in the magnetic field, the contribution of this mechanism intensifies as a result of the magnetoviscous effect.

2. The energy losses as a result of slipping of the nanoparticles and nanoaggregates in relation to the fluid matrix.

3. The mechanism of the interfacial heat exchange of the gas cavity with the wall of the container.

4. The emission of elastic oscillations into the elements of the structure and the surrounding medium.

The first mechanism is adequately interpreted on the basis of the concept of the shear wave proposed by Stokes. At high values of the diameter of the pipe d and frequency v, when $\pi d/2\lambda > 10$, because of the small depth of penetration of the viscous wave σ'' the oscillatory movement of the fluid column is of the 'piston' type, and the flow of the fluid is concentrated in the thin near-wall layer. The magnitude of the active resistance of the pipe is determined by the Helmholtz equation (10.11), and the damping factor is calculated from the expression (10.12). The contribution to energy dissipation, determined by viscous friction in the near-wall layer of the pipe within the framework of the Helmholtz model is proportional to $\sqrt{v}$.

The energy losses as a result of the 'slipping' of the particles in relation to the fluid matrix increase the shear viscosity which tends to decrease with increase of the strength of the magnetic field transverse in relation to the pipe with the magnetic fluid. However, with the increase of the number of particles in the aggregate the contribution to the dissipation of the energy of the oscillatory system, supplied by the magnetoviscous effect, is considerably greater than the contribution of the mechanism of slipping of the particles.

As a result of the adiabatic compression and tensile loading of the gas in periodic displacement of the fluid column from the equilibrium position heat exchange takes place between the gas cavity, on the one side, the walls of the pipe and the open surface of fluid, on the other side. As a result of the low heat conductivity of the gas medium the heat exchange is delayed behind the oscillations of the column. This phase shift results in some additional damping of the oscillations.

The fourth of the previously mentioned mechanisms of energy dissipation is associated with the radiation of elastic oscillations in the wall of the pipe, the auxiliary elements of the structure (holder, support), into air. Evidently, its efficiency is associated with the ratio of the acoustic resistances of the structural elements of the oscillatory system and the surrounding medium.

The last two mechanisms are not affected either by the magnetic field and are not linked with the viscosity of the magnetic fluid. The parameters capable of influencing the physical mechanisms of damping of the oscillations of the non-viscous origin – the acoustic resistance and the heat conductivity of the magnetic colloids – are almost completely independent of the strength of the magnetic field and the degree of its heterogeneity. The magnetising of the fluid cannot influence the contribution of these mechanisms to the dissipation of elastic energy of the oscillatory system.

The additional energy losses not associated with the viscosity determine the additional damping $\Delta\beta$. The experimental value of the damping factor β_{ex} consists of two components $\Delta\beta$ and β:

$$\beta_{ex} = \beta + \Delta\beta, \tag{10.25}$$

Verifying the preliminary calibration of the experimental equipment by measuring the viscosity of the investigated sample by another method ($\Delta\beta$ at selected frequency is determined), the viscosity of the fluid in the magnetic field is determined by the equation:

$$\eta = \frac{\rho d^2 (\beta_{ex} - \Delta\beta)^2}{4\pi v}. \tag{10.26}$$

Taking into account the proportionality between the effective viscosity η and the pressure gradient $\Delta p'_{1,21}$, the first of the equations (10.23) can be written in the following form:

$$\eta = \eta_0 (1 + \frac{13}{4}(\varphi - \varphi_1) + \varphi_1 L_\perp(S)). \tag{10.27}$$

Equation (10.27) contains two unknown quantities φ_1 and S, and two other parameters η_0 and φ can be directly measured. The effective viscosity, calculated from equation (10.27), can be related to the value obtained on the basis of measurements, at different combinations of φ_1 and S. The calculation results, obtained using equations (10.27), appear to be ambiguous.

For this reason it would be useful to consider the previous results, especially those obtained from the system of the equations (10.22) which gives the 'effective' value of the parameter S. Equation (10.27) can be written in the form:

$$\eta = \eta_0 \left(1 + \varphi L_\perp\right), \tag{10.28}$$

in which φ is the concentration of aggregates with the 'effective' value of parameter S.

10.3. Non-magnetic microparticles in a nanodispersed magnetic fluid

The reason for the anomalously strong dependence of the damping factor of the oscillations in the oscillatory system with the magnetic fluid inertia element on the strength of the magnetic field may be not only the presence in the disperse system of aggregates of magnetic nanoparticles, but also the presence in the system of the aggregates of the microparticles of the non-magnetic phase.

The presence in the magnetic fluid of the non-magnetic particles of the micron size predetermines the appearance of the magnetorheological effect in the system. Such a particle, placed in the magnetised medium, results in the disruption of the homogeneous (on the scale of the particle) distribution of the strength of the magnetic field. The field at the poles of the particles weakens and at the equator it becomes stronger. Therefore, the adjacent particles in the equatorial region will be repulsed and those along the polar axis will attract each other, trying to occupy a position in the area of the minimum value of the strength of the magnetic field. The increase of the effective viscosity in the field is determined by the processes of formation and failure in the medium of non-magnetic structures as a result of the competing effect of the magnetic and hydrodynamic forces.

B.E. Kashevskii, V.I. Kordonskii and I.V.Prokhorov (1988) derived the expressions for the increase of the viscous stresses in a suspension of non-interacting ellipsoids [43]:

$$\Delta\tau_{max} = \frac{1}{2}\mu_0 M^2 \varphi'$$
(10.29)

and for the characteristic duration of orientation relaxation of the internal structure of the medium:

$$t' = \eta_0 \lambda^2 / \mu_0 M^2 \ln\lambda,$$
(10.30)

In these equations, η_0 is the viscosity of the disperse medium, φ' is the concentration of the non-magnetic phase, λ is the ratio of the major and minor half-axes, M is magnetisation. The orientation time t' determines the shear rate $\gamma = 1/t'$ up to which the viscous stresses in the magnetic field rapidly increase. Using the Newton law $\Delta\eta = \Delta\tau_{max}/\dot\gamma$ and equations (10.29) and (10.30), we can derive the expression for the increase of the viscosity in the magnetic field associated with the presence of the non-magnetic particles:

$$\Delta\eta_{max} = \frac{1}{2}\varphi'\eta_0 \frac{\lambda^2}{\ln\lambda}.$$
(10.31)

Questions for chapter 10

* *Name the components of the oscillatory system with the inertia element in the form of a magnetic fluid column,*
* *Name the physical mechanisms of energy dissipation in the magnetic fluid column carrying out reciprocating flows in the absence of the magnetic field?*
* *Describe the conditions under which we can use the equation for the damping factor of the oscillations of the magnetic fluid column derived from the Poiseuille flow model?*
* *Describe the conditions for using the equation for the damping factor of the oscillations of the magnetic fluid column derived from the Helmholtz flow model.*
* *Explain the anisotropy of the properties of the magnetic fluid in the model constructed by V.A. Naletova.*
* *Name the physical mechanisms of dissipation of elastic energy in the oscillatory system with the inertia element in the form of a magnetic fluid column. Compare their contribution.*
* *Why is the damping of the oscillations influence by the presence of the non-magnetic particles in the magnetic disperse system?*
* *Describe the method for the determination of the form of the non-magnetic particles, dispersed in a nanodispersed magnetic fluid.*
*

11

Kinetic and strength properties of a magnetic fluid membrane

At present, the magnetic fluids are used mostly in magnetic fluid seals (MFS) and also magnetic fluid gaskets (MFG), employed in different areas of engineering, including aerospace, electrical engineering and biotechnological branches. Of considerable importance are the strength and kinetic properties of these devices. However, the investigation of these devices is a complicated experimental task. Therefore, it is interesting to investigate the magnetic fluid membrane which may be used as a model of the MFS and MFG [3].

The magnetic fluid membrane is a droplet of a magnetic colloid overlapping the cross-section of a pipe with the inner diameter of ~1.5 cm as a result of the stabilising effect of the heterogeneous magnetic field. In the presence of a bottom in the pipe, the magnetic fluid breach isolates the air cavity below it. In this case, the magnetic fluid functions as an incompressible medium, and its properties, such as the magnetic control of the free surface, fluidity, inertia, become very important.

Because the shape of the surface of the magnetic fluid droplet in the absence of gravitation and capillary forces is determined by the parameters of the magnetic field, the forced rupture of the membrane (for example, as a result of creating a pressure drop) is followed immediately by the restoration of its integrity. Consequently, in contrast to the conventional fluid films, the magnetic fluid membrane is capable of self-restoration.

The results of measurements of the critical pressure drop p_k, resulting in the rupture of the magnetic fluid membrane, will be discussed. To produce a magnetic fluid membrane in this case, experiments were carried out with a glass pipe (350 mm long) with a flat bottom, the inner diameter of the pipe is 13.5 mm. In the experiments, the membrane was formed by the method of 'self capture' of a portion of a magnetic fluid by a ring-shaped magnet introduced through the bottom of the pipe containing the colloid, and lifted to a height of h above the level of the fluid. The ring-shaped magnet was connected with the kinematic unit of the cathetometer.

Experiments were carried out using the magnetic fluids employed in engineering in the form of a colloidal solution of single-domain particles of magnetite Fe_3O_4 in kerosene (MF-1 and MF-2) and in organic silicon (MF-3). The physical parameters of the magnetic fluid are presented in Table 11.1.

Table 11.1.

Sample	ρ, kg/m^3	η_s, Pa·s	M_S, kA/m	χ
MF-1	1294	$3.2 \cdot 10^{-3}$	52 ± 1	6.2
MF-2	1499	$8.1 \cdot 10^{-3}$	60 ± 1	7.5
MF-3	1424		43 ± 1	5.0

Figure 11.1 shows graphically the results of measurement of the distance h_k between two consecutive ruptures of the membrane in dependence on the height of the air column h isolated by the fluid for the specimens: a – MF-1; b – MF-2; c – MF-3.

The density of the fluid was measured using a pycnometer, viscosity by the capillary method. The saturation magnetisation of the colloid M_S was determined by extrapolating the dependence $M = f(H^{-1})$ to the range of strong magnetic fields. Magnetic susceptibility was determined from the slope of the tangent to the curve $M(H)$ in the initial section.

The individual results obtained for h_k cannot be used to determine the tendency for the variation of this parameter with increase of the height of the air column. The form of this dependence can be determined only by averaging a large number of experimental data (no less than 50 for MF-1, 150 for MF-2) in a narrow range of the displacement of the magnetic head from h to $h + \Delta h$ ($\Delta h \approx 1$ cm) for several greatly different values of the height of the air column h. Consequently, the average value $\bar{h}_k$ was obtained.

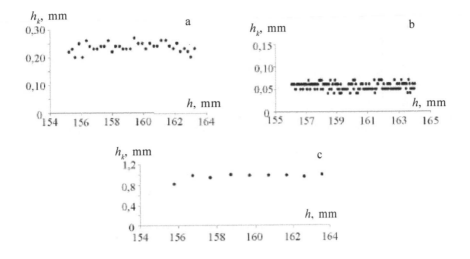

Fig. 11.1. Dependence $h_k(h)$: 1– MF-1; 2 – MF-2; 3 – MF-3.

The ruptures of the magnetic fluid membranes which were not observed when moving the magnetic head in the reverse direction within the range of limited width 2Γ, where Γ is the distance between the initial equilibrium position and the first displace equilibrium counted on the cathetometer. For MF-1, MF-2 and MF-3 we obtain: $\Gamma_1 = 1.77$, $\Gamma_2 = 3.04$ and $\Gamma_3 = 4.53$ (mm).

Figure 11.2 shows in the coordinates $P(z)$ the model of the thermodynamic process taking place in the gas cavity assuming slow lifting of the magnetic head along the pipe with a constant cross-section.

In the initial position of the membrane the height of the air column is h_0 and the pressure in the gas cavity is $p_a = p_0 + \rho gb$, where p_0 is the external (atmospheric) pressure.

As an example, we use the following values: $b = 1.3$ cm, $p_0 = 10^5$ Pa, $\rho = 1294$ kg/m³ and $\rho gb = 165$ Pa.

In this situation, the membrane is in the region with the maximum magnetic field, and the displacement of the magnetic head both upwards and down to the first rupture of the membrane is the same and equal to Γ, and

$$\Gamma = h_g + h_d, \qquad (11.1)$$

where h_g is the increase of the height of the air column; h_d is the displacement of the membrane in relation to the magnetic head.

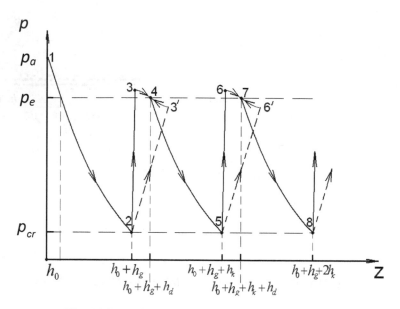

Fig. 11.2. Model of the thermodynamic process.

In the sections 1–2, 4–5 and 7–8 the isothermal expansion of the gas cavity takes place.

The critical pressure drop, causing the rupture of the membrane during its displacement from the initial equilibrium position to the first fracture is:

$$p'_{cr} \equiv p_a - p_{cr}, \qquad (11.2)$$

where p_{cr} is the pressure in the gas cavity at which the membrane ruptures. In isothermal expansion of the gas cavity

$$p_a h = p_{cr}(h + h_g), \qquad (11.3)$$

from which

$$p_{cr} = p_a \frac{h}{h + h_g}. \qquad (11.4)$$

Because of the third Newton law

$$h_g k_g = h_d k_p. \qquad (11.5)$$

Here k_g is the isothermal elasticity coefficient of the gas cavity $k_g = \dfrac{p_a S}{h}$; coefficients k_g and k_p are constant.

Taking into account (11.1), we obtain

$$h_g = \Gamma \frac{k_p}{k_p + k_g}.$$ (11.6)

After substitution of (11.6) into (11.4) we obtain

$$p_{cr} = p_a \frac{h}{h + \tilde{\gamma}\Gamma}.$$ (11.7)

where $\tilde{\gamma} \equiv \dfrac{k_p}{k_p + k_g}$.

Substituting (11.7) into (11.2) gives

$$p_{cr_1} = \frac{\tilde{\gamma} p_a \Gamma}{h + \tilde{\gamma}\Gamma}.$$ (11.8)

For the investigated samples MF-1, MF-2 and MF-3, the values of p_{k_1} are equal to $5.7 \cdot 10^2$ Pa, $10.4 \cdot 10^2$ Pa and $12.4 \cdot 10^2$ Pa, respectively.

The experiments with the determination of p_{cr_1} by the hydrostatic method results in the lower values of these parameters (~30% lower). The difference between the experimental results is explained by the dependence of the elasticity of the ponderomotive type on the special features of the viscous flow of the magnetic fluid during displacement of the membrane at different rates of application of the pressure, and also by its non-linear form. Comparison of the results makes it possible to estimate the static values of parameters $\tilde{\gamma}$ and k_p.

In the conditions 2, 5 and 8 (Fig. 1.2), the integrity of the membrane is disrupted as a result of the formation of a circular orifice in its central part. Air is supplied into the orifice under the effect of the pressure drop and this results in a jump-like increase of pressure. In this stage of the process, the membrane is displaced in the direction of the equilibrium position, i.e. in the direction ∇H, which, on the one hand, results in some increase of the volume of the gas cavity and, on the other hand, creates suitable conditions for closure of the cavity.

The variation of the condition of the gas in the gas cavity during the existence of the orifice ('lifetime' of the orifice τ) may develop by one of the two variants of the transition to a new state with the equilibrium pressure p_c, each of which results in the excitation of intrinsic oscillations of the magnetic fluid membrane.

In the first variant (transitions 2–3, 5–6, etc) the displacement of the membrane is small because of its inertia and existence of the 'rigid connection' of the restoration of integrity with the topography of the magnetic field, and the increase of pressure in the cavity is caused by pumping of air through the orifice taking place by the adiabatic mechanism.

In the second variant (transitions 2–3'; 5–6', etc, indicated by the dashed line in Fig. 11.2) the membrane passes during time τ the equilibrium position and at the moment of closure of the orifice (points 3'; 6') interrupts its movement and then travels in the reverse direction to the equilibrium position. This variant is hypothetically possible if there is no 'rigid' connection of the integrity of the membrane with the topography of the magnetic field, and the process of slowing down the displacement and closing of the membrane is determined mainly by the gas dynamic effect of the increase of the force of resistance to the movement of the gas flow with increase of its speed.

After completing the damping oscillations and the establishment of the thermodynamic equilibrium, the gas in the cavity is in the state indicated by the points 4 and 7 in the graph. With a further lifting of the magnetic head the fracture of the membrane takes place at a small increase of the pressure in the gas cavity.

We derive the equation for calculating the increment of the pressure in the gas cavity in subsequent ruptures of the membrane p'_{cr}:

$$p'_{cr} = p_e - p_{cr}$$
(11.9)

At the isothermal expansion of the gas

$$p_e = p_a \frac{h}{h + h_g + h_d - h_k}.$$
(11.10)

where h_d is the displacement of the membrane in relation to the magnetic head.

In the given stage of expansion of the gas cavity

$$h_{cr} = h'_g + h_d.$$
(11.11)

where h'_g is the increase of the height of the gas cavity.

Substituting (11.7) for p_{cr} into (11.9) and also expression (11.10) for p_c, and also taking into account the equilibrium condition of the forces $h'_g k_g = h_d k_d$, we obtain

$$p'_{cr} = p_a h \frac{h_{cr} + \tilde{\gamma}\Gamma - h_g - h_d}{\left(h + h_g + h_d - h_{cr}\right)\left(h + \tilde{\gamma}\Gamma\right)}. \tag{11.12}$$

Taking into account (11.2) and the relationship

$$\frac{h_d}{h_{cr}} = \frac{k_g}{k_g + k_p}, \tag{11.13}$$

the equation (11.10) can be written in the form

$$p'_{cr} = p_a \frac{\tilde{\gamma} h h_{cr}}{\left(h + \tilde{\gamma}\Gamma - \tilde{\gamma}h_{cr}\right)\left(h + \tilde{\gamma}\Gamma\right)}. \tag{11.14}$$

Taking into account the above notations, the expression for p_p (11.8) can be presented in the following form

$$p_e = p_a \frac{h}{h + \tilde{\gamma}\left(\Gamma - h_{cr}\right)}. \tag{11.15}$$

All the quantities included in the equations (11.14) and (11.15), are determined directly or indirectly from the experimental data.

In the experiment conditions $h \gg \tilde{\gamma}\Gamma$ and, therefore, in this case the equation (11.14) can be simplified

$$p'_{cr} \cong p_a \tilde{\gamma} \frac{h_{cr}}{h}. \tag{11.16}$$

When comparing the expressions (11.8) and (11.16) we obtain $p_{cr1} \gg p'_{cr}$, since $\Gamma \gg h_{cr}$.

There is a small increase of p'_{cr} with increase of the height of the gas cavity h for the MF-1 and MF-3. A larger decrease of p'_{cr}, obtained for the MF-2 sample, is evidently associated with the decrease of the mass of the membrane as a result of the consumption of its mass on the internal surface of the pipe, as observed by visual examination.

The magnetic fluid membrane, prepared using the MF-1 colloid, characterised by a lower concentration of the magnetic phase and, correspondingly, smaller values of the parameters M_S and χ in comparison with the other colloids with the identical carrier MF-2, is characterised nevertheless by a considerably higher value of the increment of pressure p'_{cr} (although the relationship for the parameter

p'_{cr} is reversed) which at first sight appears to be unexpected. The physical nature of the result is relate d to the higher integrity of the membrane based on the magnetic fluids with a high value of χ by the given the topography of the magnetic field. The restoration of its integrity takes place at smaller displacement in the direction of ∇H.

The mass of the gas portion Δm, transmitted by the magnetic fluid membrane during a single rupture, within the framework of the expected thermodynamic process (Fig. 11.2) is obtained from the equation of state of the ideal gas written for two adjacent states

$$p_e V = \frac{m}{\mu} RT \text{ and } p_e \left(V + \frac{\pi d^2}{4} \overline{h}_{cr} \right) = \frac{m + \Delta m}{\mu} RT, \qquad (11.17)$$

where μ is the molar mass of the gas; R is the universal gas constant, T is absolute temperature.

From the system of equations we directly obtain:

$$\Delta m = \frac{\mu p_e \pi d^2 \overline{h}_{cr}}{4RT}. \qquad (11.18)$$

Equation (11.15) at h $\gg \tilde{\gamma}(\Gamma - h_{cr})$ gives $p_e \approx p_a$.

Accepting for air $\mu = 30$ g/mole, $p_e = 10^5$ Pa, $d = 1.36 \cdot 10^{-2}$ m, $T = 298$ K, we obtain $\Delta m = 1.76 \cdot 10^{-4} \overline{h}_{cr}$.

The smallest value $\Delta m_{min} = 0.009$ mg is obtained for the magnetic fluid membrane based on MF-2, the highest value $\Delta m_{max} = 0.17$ mg is characteristic of the magnetic fluid membrane based on MF-3. Regulating the amount of the colloid introduced into the membrane, it is possible to expand slightly the range of the values of Δm. For example, Δm decreases greatly as a result of minimising the amount of MF-2 in the membrane. In this case, in the absence of the special measures for thermostatic control, vibrational and acoustic insulation, the process of rupture–restoration of the magnetic membrane may become uncontrollable.

The speed of the airflow through the orifice is estimated on the basis of the relationship linking the pressure drop in the orifice Δp_g and the speed in the area of the maximum compression of the flow v_g [44]:

$$\Delta p_g = \frac{1}{2} \rho_g v_g^2 \xi \left[\frac{\sigma}{S} \right], \qquad (11.19)$$

where σ is the area of the orifice; S is the cross-sectional area of the pipe; ξ is the hydraulic resistance coefficient which depends on the area of the orifice σ and the Reynolds number.

For the case $\sigma \ll S$, $\xi = \xi_0 = 2.9$, we can write:

$$V_g = \sqrt{\frac{2\Delta p_g}{\rho_g \xi_0}}. \tag{11.20}$$

The pressure drop

$$\Delta p_g = p_a - \frac{p_e + p_{cr}}{2} = \frac{1}{2}(p_a - p_e + p_a - p_{cr}). \tag{11.21}$$

Taking into account the previously derived equations (11.7) and (11.15), we carry out elementary transformations

$$\Delta p_g = \frac{p_a}{2}\left(1 - \frac{h}{h + \tilde{\gamma}(\Gamma - h_{cr})} + 1\frac{h}{h + \tilde{\gamma}\Gamma}\right). \tag{11.22}$$

since $\Gamma \gg h_{cr}$, then at $h \gg \tilde{\gamma}\Gamma$ we can write:

$$\Delta p_g \cong \frac{\tilde{\gamma}\Gamma}{h}p_a. \tag{11.23}$$

Substituting (11.23) into (11.20) we obtain

$$V_g \approx \sqrt{\frac{2\tilde{\gamma}p_a\Gamma}{\rho_g \xi_0 h}}. \tag{11.24}$$

The 'lifetime' of the orifice τ can be estimated from the equation:

$$\tau = \frac{\Delta m}{\rho_g \sigma v_g}. \tag{11.25}$$

The direct measurement of σ was not carried out in this experiment. To carry out the approximate calculations, it may be assumed that the diameter of the orifice is in the range 0.1÷0.3 cm. Estimating the upper value of τ, we carry out calculations using the smallest of the given values.

The lifetime of the orifice is shorter than the period of oscillations of the membrane (10÷15 ms) so that the first variant of the transition of the magnetic fluid membrane to the equilibrium state (Fig. 11.2) can be regarded as more probable. The argument favouring this conclusion is the fact that the increment of the potential energy of the oscillatory system at the moment of rupture of the membrane, calculated using the equation

$$\Delta E_p = 0.5\left(k_p h_d^2 + k_g h_g'^2\right),\qquad (11.26)$$

is equal to (in the experiments with MF-1) only $2 \cdot 10^{-6}$ J, whereas the kinetic energy, calculated from the average speed of displacement of the membrane (according to the second variant) would be:

$$E_k \geq \frac{\pi \rho d^2 b h_d^2}{4\tau^2} \approx 2.6 \cdot 10^{-5}\,\text{J}.\qquad (11.27)$$

Questions for chapter 11

• *What is the magnetic fluid membrane? Which devices can be modelled using the magnetic fluid membrane?*
• *What is the main difference between the magnetic fluid membrane and the conventional fluid films?*
• *Describe the experimental equipment for studying the process of rapture and restoration of the magnetic fluid membrane.*
• *What are the physical processes taking place during the rapture and restoration of the magnetic fluid membrane.*
• *What is the critical pressure drop causing the rapture of the membrane? Is this value constant for identical experiment conditions?*
• *How does the process of rapture and restoration of the magnetic fluid breach depend on the properties of the magnetic fluid?*
• *Determine the lifetime of the orifice at rapture of a magnetic fluid membrane.*

12

The cavitation model of the rupture–restoration of the magnetic fluid membrane

In the previous chapter, attention was given to the method and results of measurement of the kinetics and strength parameters of the magnetic fluid membrane. However, it is necessary to explain the physical nature of factors which determine the processes of rupture and restoration of the orifice and predetermine the numerical values of these parameters.

The exact analytical solution of this problem, based on the application of the magnetohydrodynamics equations taking into account the surface tension forces, the specific geometry of the magnetic field and the free surface of the magnetic fluid membrane, is a very complicated task. It is therefore interesting to consider the results of extensively examined (theoretically and experimentally) problems having the same physical characteristics as the task we are facing here. In this respect, the model of rupture–restoration of the magnetic fluid membrane, based on the results of investigations of acoustic cavitation [45, 46] is highly promising. Investigations of the acoustic and hydrodynamic cavitation started with the work of Rayleigh (1917) [47] who examined the processes of closure of a spherical (not filled with gas) cavity in an ideal fluid.

The diagram of the cavitation model of the process of rupture–restoration of the magnetic fluid membrane is shown in Fig. 12.1. The black solid crosshatching shows the magnetic fluid membrane with a circular orifice with a radius R in the centre. The upper and lower open surfaces of the membrane have the form of a conical

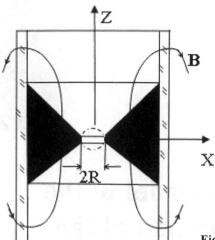

Fig. 12.1. Diagram of the cavitation model.

surface with the Z axis. The magnetic field with induction **B** is generated by a ring-shaped magnet, magnetising in the axial direction and situated coaxially in relation to the axis of a glass pipe in the plane of symmetry of the magnetic fluid membrane (not shown on the figure). The simulated cavitation cavity is indicated by the dot–dash line. Thus, for transition to the magnetic fluid membrane from the 'standard' spherically symmetric scheme of closure of the cavitation cavity we exclude two spherical sectors (at the top and bottom).

A specific feature of the proposed model is the assumption that the function of the hydrostatic pressure in this case is fulfilled by the pressure of the ponderomotive origin on the side of the heterogeneous magnetic field of the ring-shaped magnet. This assumption is made on the basis of the fact that the orifice in the magnetic fluid membrane in contrast to the cavitation cavity is not a closed surface. It is also assumed that the effect of ponderomotive forces on both surfaces of the magnetic fluid membrane results in the distribution of pressure in the fluid and the equivalent distribution of pressure at the spherically symmetric flow. In the investigated model, as in the formulation of the Rayleigh problem, no account is made of the effect of surface tension forces and gravitational force.

To evaluate the ponderomotive effect of the magnetic field on the magnetic fluid membrane, experiments were carried out to examine the topography of the magnetic field of a ring-shaped magnet used in the experiments. In particular, the distribution of the strength and strength gradient of the magnetic field at the points of the active zone (zone in which the magnetic fluid membrane is located) was determined.

Taking into account the topography of the magnetic field with $M_z \gg M_r$, because of the symmetry of the magnetic fields in relation to the plane $Z = 0$ we obtain the following expression for the axial component of the ponderomotive force by analogy with equation (5.84)

$$\Delta f_z = -2\mu_0 S \left(M_z \frac{\partial H_z}{\partial z} \right)_{z=-\frac{b}{2}} \delta z,$$

where δz is the displacement of the membrane in relation to the magnet after rupture and escape of a portion of air ($\delta z \ll b$), i.e. $\delta z = h_d$.

Correspondingly, the estimate of the excess pressure on the magnetic fluid membrane p_0 can be obtained from the expression:

$$p_0 = -2\mu_0 \left(M_z \frac{\partial H_z}{\partial z} \right) h_d. \tag{12.1}$$

The equation cited most frequently in the cavitation literature is the equation of motion of the surface of the solitary gas-filled spherical cavity in an infinite fluid with the density ρ_0:

$$R\frac{d^2 R}{dt^2} + \frac{3}{2}\left(\frac{dR}{dt} \right)^2 + \frac{1}{\rho_0}[p_\infty - p(R)] = 0, \tag{12.2}$$

where R is the actual radius of the cavity; t is time; p_∞ is the pressure in the fluid at infinity; $p(R)$ is the pressure on the cavity surface.

Using the initial conditions according to which if $t = 0$ then $R = R_0$ and the rate of expansion of the cavity $U = 0$, integration of (12.2) at $p_\infty = p_0$ (p_0 is hydrostatic pressure) leads to the expression

$$U^2 = \frac{2}{3}\frac{p_0}{\rho_0}\left(1 - \frac{R_0^3}{R^3} \right), \tag{12.3}$$

where ρ_0 is the density of the fluid.

The maximum value of the rate of expansion of the cavity U_{max} is obtained at $R \to \infty$:

$$U_{max} = \sqrt{\frac{2}{3}\frac{p_0}{\rho_0}}. \tag{12.4}$$

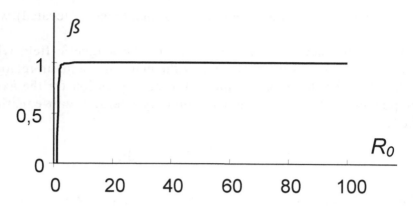

Fig. 12.2. The speed of the boundary of the expanding bubble at a constant tensile stress.

Equation (12.4) contains an important result according to which the rate of expansion of the sphere at the initial period of movement rapidly increases, approaching its asymptotic value.

Figure 12.2 shows the graph of the dependence of the relative speed $\beta = U/U_{max}$ on the radius expressed in the units of R_0. If, for example, the initial radius (the radius of the cavitation nucleus) is $R_0 = 10$ μm, and the maximum radius is $R_m = 1$ mm, then, as indicated by Fig. 12.2, the average value of the rate of expansion of the spherical cavity is $\bar{U} \approx U_{max}$.

Rayleigh investigated the case of a hollow cavity $[p(R) = 0]$ and constant pressure at infinity, when $p_\infty = p_0$. Under these conditions, equation (12.2) gives the rate of closure of the cavity U':

$$U'^2 = \frac{2}{3}\frac{p_0}{\rho_0}\left(\frac{R_m^3}{R^3} - 1\right),$$ (12.5)

where R_m is the initial (maximum) radius of the cavity; p_0 is hydrostatic pressure.

Integration of (12.5) using the Γ-functions leads to the well-known Rayleigh equation for the closure time of a hollow cavity in the field of hydrostatic pressure:

$$\tau_m = 0.915 \cdot R_m \cdot \left(\rho_0 / p_0\right)^{\frac{1}{2}}.$$ (12.6)

Taking into account the experimental data, we assume that $M_S = 45$ kA/m, $h_d = 0.1$ mm, $\partial H_z/\partial z = 4.6 \cdot 10^6$ A/m^2. After substituting these values into (12.1) we obtain $p_0 \approx 50$ Pa.

The estimate of the closure time of the orifice using equation (12.6) with substitution of the resultant value of p_0 and also $R_m = 10^{-3}$ m, $\rho_0 = 10^3$ kg/m^3 gives $\tau_m \approx 4.5$ ms, which is in satisfactory agreement with the experimental data. The rate of expansion of the cavity U_{max}, calculated from equation (12.4), is 0.18 m/s.

These examples show that the results of experimental examination of the process of expansion–closure of the orifice in the magnetic fluid membrane and of the calculations, based on the cavitation model, are in good agreement with each other.

The rupture strength of the fluid is determined by the presence in the fluid of cavitation nuclei and their distribution in the volume. It is therefore interesting to examine the processes of displacement of microscopic gas bubbles under the effect of ponderomotive forces.

The air bubble is subjected by the following force from the side of the heterogeneous magnetic field in the magnetic fluid:

$$F_b = -4\pi\mu_0 M\nabla H \cdot R_b^3 / 3,$$

where the R_b is the radius of the cavitation nucleus.

In the plane $Z = 0$ the bubble is located in a heterogeneous magnetic field with the radial component $\Delta H_z/\Delta r$ and, consequently, the bubble moves in the viscous fluid to the centre of symmetry of the magnetic field (into the region with a lower strength of the magnetic field) with the average velocity:

$$\overline{v} = 2\mu_0 M\overline{\nabla H}R_b^2 / 9\eta.$$

The bubble travels from the wall to the centre during the time

$$\overline{\tau} = \frac{d}{2\overline{v}} = \frac{9d \cdot \eta}{4\mu_0 M\overline{\nabla H}R_b^2}$$

The radius of the neck of the bubble r_b in separation of the bubble from the wall is calculated from the equation

$$r_b = 2\mu_0 M\nabla H R_b^3 / 3\sigma,$$

where σ is the surface tension coefficient of the magnetic fluid.

Assuming that $\overline{\nabla H} = 3.91 \cdot 10^6$ A/m^2, $M = 40 \cdot 10^3$ A/m, $\eta = 4.7 \cdot 10^{-3}$ Pa·s, $\sigma = 28 \cdot 10^{-3}$ N/m, $R_b = 10$ µm, we obtain $\overline{v} = 9.3 \cdot 10^{-4}$ m/s, $\overline{\tau} = 7$ s, $r_b = 4.7 \cdot 10^{-3}$ µm.

Therefore, the central part of the magnetic fluid membrane, which is the thinnest part of the magnetic fluid membrane, as a result of flow of the fluid into the region near the wall (the region with the highest strength of the magnetic field), is characterised by the minimum mechanical strength as a result of the buildup of cavitation nuclei.

Questions for chapter 12

• *Explain the term 'the hydrodynamic cavitation'. Who described this phenomenon theoretically for the first time? Where does it occur in practice?*
• *Draw the diagram of the cavitation model of rupture of the magnetic fluid membrane. Name the special features of this model?*
• *Write the equation for the rate of expansion of the gas cavity, analyse the equation.*
• *How is the strength of the magnetic fluid membrane linked with the presence of cavitation nuclei in it?*
• *Why is the central part of the magnetic fluid membrane characterised by the lowest strength?*

13

Methods for producing magnetic
fluids and ferrosuspensions

The methods of producing magnetic fluids can be used to produce
different fluids with the required properties [50]. The technologies
of producing magnetic fluids have been extensively developed and
can be used to organise the mass production of relatively cheap
magnetic fluids with the required properties. Magnetic fluids with
low or high evaporability have been developed; other fluids are
capable of wetting the surface and absorbing moisture from the
air; non-toxic fluids are used for medical purposes and electrical
conducting fluids for electromechanical systems. This large variety
offers considerable possibilities for using magnetic fluids in technical
devices and processes.

The composition of a magnetic fluid must contain three
components: a fluid base (or, in other words, the liquid carrier),
magnetic particles of colloidal dimensions (magnetics), and a
stabiliser preventing bonding of the colloidal particles. Each
component should satisfy certain requirements and only if this
condition is satisfied it is possible to produce a magnetic fluid
suitable for use in a specific device [4].

The fluid base is selected depending on the application. For
example, in lubricating and sealing systems it is necessary to use
magnetic fluids based on mineral oils and organic silicon compounds,
separation systems and printing devices use fluids based on
hydrocarbons, and water-based fluids are used in medicine. The fluid
basis subject to requirements to ensure low evaporation, no toxicity,
resistance in corrosive media, insolubility in specific media, etc.

The size of the magnetic particles should be sufficiently small because the stability of the magnetic fluid as a colloidal system is ensured by the thermal motion of the particles preventing their bonding and settling, and the speed of their motion increases with a decrease of the particle size. At the same time, the particles should not be too small because if they are smaller than 1–2 nm they lose their magnetic properties and convert to the paramagnetic state. The material of the particles should be characterised by high magnetisation and the capacity to produce small particles with a small dispersion of the dimensions.

In coarse-disperse systems with the particles larger than 1 μm the Brownian motion is not so active and in the suspensions with a high viscosity of the disperse medium this motion does not occur. In this case, the controlling role is played by the settling rate of the particles under the effect of gravitational force (*sedimentation* rate). The sedimentation rate is characterised by the Stokes law:

$$v = \frac{2r^2 g}{9\eta}(\rho_2 - \rho_1),$$

where r is the particle radius, ρ_1 is the density of the liquid carrier, ρ_2 is the particle density, g is the gravitational acceleration, η is the viscosity of the disperse system.

The sedimentation resistance of the ferrosuspensions depends on the concentration of the disperse phase. It is ensured by both the direct structurisation of the system as a result of the action between the particles of the molecular forces and by adding additional surface-active substances (SAS, surfactants) which form the structural–mechanical surfaces between the particles.

All the disperse systems (both micro- and nano-) are conventionally divided into two groups on the basis of the coefficient of interfacial surface tension σ which can be greater or smaller than some critical value σ_c:

$$\sigma_c = \gamma * \frac{k_0 T}{\delta^2},$$

were $\gamma*$ is a dimensional multiplier ($\gamma* \sim 30$), δ is the average particle size, T is absolute temperature, k_0 is the Boltzmann constant. At room temperature $\sigma_c \sim 10^{-4}$ N/m.

The first group includes *lyophilic* disperse systems characterised by low values of the interfacial surface energy ($\sigma < \sigma_c$). These systems form spontaneously by colloidal dissolution which is accompanied by the increase of entropy because of the more uniform distribution of the disperse phase particles.

The second group includes *lyophobic* disperse systems characterised by high interfacial tension ($\sigma > \sigma_c$) and, consequently, a distinctive interface. The molecular forces on the surface of the lyophobic particles cause coalescence (bonding) – the process leading to the aggregate instability of such thermodynamically non-equilibrium systems. The aggregate stability of the lyophobic disperse systems is achieved by formation on the particles of adsorption–solvate layers which prevent particle coalescence.

This problem is solved using different stabilisers which include substances (polymers or SAS) with long-chain molecules, containing functional groups (OOH, H_2OH, H_2NH_2, etc). The stabiliser is selected to ensure that its functional groups interact with the substance of the magnetic particle, forming a monomolecular shell strongly bonded with the particle. The long-chain part of the stabiliser molecule should be compatible with the liquid carrier in order to take part in thermal motion and, consequently, prevent convergence of the particles. The selection of the stabiliser is a relatively complicated task solved mostly by experiments.

In non-polar disperse media (oil, kerosene, dodecane, octane, etc) the flexible non-polar ends of the SAS, related to the liquid carrier, are directed from the particle to the fluid, Fig. 13.1a. The stability of the disperse particles in the polar fluid, for example, in water, is achieved by the characteristic distribution of the two SAS layers (Fig. 13.1b), and the polar ends of the second layer of the SAS, related to the liquid carrier, are directed from the particle to the fluid. In this case, the thickness of the shielding shell is twice the thickness of the shielding shell of, for example, magnetite, stabilised in hydrocarbon media.

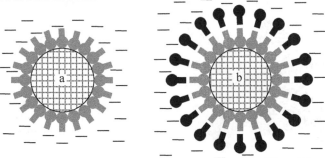

Fig. 13.1. Stabilised magnetic nanoparticles.

Varying the composition of the magnetic fluids, i.e., selecting the liquid carrier, the magnetic, stabiliser, results in the formation of fluids with the required properties.

13.1. The production of magnetic fluids with different disperse phases

To obtain such a complicated medium such as magnetic fluids, it is necessary to solve a number of problems. In particular, it is necessary to produce magnetic particles with the size not larger than 5–10 nm. It is also important to coat the particles with a layer of the stabiliser molecules. Finally, the stabiliser should not only prevent bonding of the particles but also ensure the formation of a colloidal solution in the liquid carrier. The combined solution of these problems produces a magnetic fluid stable in the gravitational, centrifugal and magnetic fields and characterised by high magnetic properties and behaving in many aspects as a continuous homogeneous medium [4]. The particles of the required dimensions can be produced by different methods.

Crushing method
The first magnetic fluid was produced in the middle of the 60^s by milling magnetite in a ball mill in a solution of oleic acid in kerosene. Around three months of continuous operation of the mill were required to produce a stable colloidal system with excellent magnetic properties. The particles of the colloidal dimensions are produced in the ball mill as a result of milling magnetite with steels balls rolling in a horizontal rotating cylinder, Fig. 13.2.

The optimum filling of the meal with a spheres is 30–40% of its volume, and 20% with the disperse material. Milling in the presence of the liquid carrier and the stabiliser produces sufficiently fine

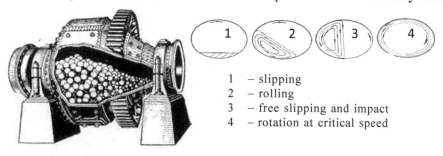

1 – slipping
2 – rolling
3 – free slipping and impact
4 – rotation at critical speed

Fig. 13.2. The ball mill: general view and working conditions.

particles and also ensures colloidal stability of the system. In dry milling it is usually not possible to produce particles smaller than 60 nm.

The production of the magnetic fluids in ball mills has both advantages and shortcomings. The method is very simple to apply and produces particles with the required dispersion as a result of controlling the milling time; there are no losses of the solvent if the solvent is volatile; the method can be used to produce fluids with different magnetics (such as magnetite, cobalt, nickel, iron, ferrites, etc), using different bases (kerosene, hydrocarbons, water, silicones, organic fluorine components, etc). However, the rubbing spheres contaminate the produced magnetic fluid and the dimensions of the resultant particles greatly differ and this has a negative effect on the colloidal stability of the fluid.

The main shortcoming of the method is the long duration of the process and the small yield of the final product; in operation of a mill for several weeks only 200–300 ml of the magnetic fluid is produced.

In addition to this, the magnetic characteristics of the pool was fluids are relatively low (saturation magnetisation approximately 10 kA/m) and to increase them the fluid is evaporated and is additionally decreases productivity of the method. At the same time, the size of the particles varies over a wide range resulting in an ambiguous relationship between the properties of the magnetic fluids and their composition.

The need to reduce the price of magnetic fluids has resulted in attempts to improve the milling method. For example, it has been attempted to replace magnetite by a more brittle non-magnetic ferrous oxide – wüstite FeO. After milling, the produced non-magnetic colloid is heated and, consequently, the colloid breaks up to magnetite and α-iron according to the equation:

$$4FeO \rightarrow Fe + Fe_3O_4.$$

This method is used to produce fluids with approximately the same characteristic (for example, saturation magnetisation approximately 10 kA/m) as in direct milling, but at a considerably higher rate. In addition to mechanical crushing, it is also possible to use other methods of breaking up the materials for producing fine-dispersion magnetic particles. For example, attempts have been made to use ultrasound and electric plasma refining and also refining with a

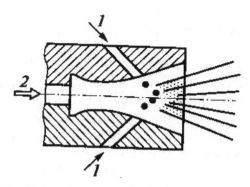

Fig. 13.3. A nozzle for spraying: 1) the melt; 2) gas.

rotating magnetic field or electric sputtering (Fig. 13.3). However, these attempts have not improved the characteristics of the magnetic fluids nor improved the productivity of the process. In addition, the crushing methods can be used only for brittle magnetics, such as ferrites or magnetite, where magnetisation is not very high. It is hardly possible to use these methods for producing colloidal particles of the metals whose magnetisation is several times higher, because of the high plasticity of the metals.

Condensation methods
The particles of the colloidal sizes can be produced as a result of joining (condensation) of the individual molecules. In joining the molecules or atoms, the free energy of the system decreases and, therefore, the process is spontaneous. The size of the resultant particles is strongly affected by the conditions in which the individual molecules join into particles and, therefore, various variants of the method are used to produce the colloidal particles of the magnetics.

Historically, the production of high-dispersion magnetics by the condensation methods was used for the first time for metal particles, i.e., to produce magnetic fluids with high magnetic properties. One of the first methods to be used was the *carbonyl method* based on the thermal dissociation of metal carbonyls.

The vapours of the metal carbonyl, produced by evaporation in a container 1 (Fig. 13.4), nd diluted by an inert gas (for example, nitrogen or argon), travel into the reactor 2 where the dissociation of the metal pentacarbonyl takes place at elevated temperatures. The metal atoms merge into particles and the volatile products of dissociation condense in equipment 3. The metal particles with the size of 2–30 nm can be produced by changing the conditions in the

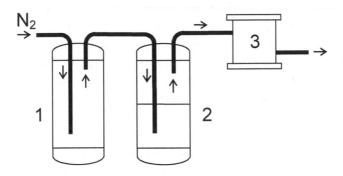

Fig. 13.4. Diagram of producing the magnetic fluid by dissociation of metal carbonyls.

reactor (temperature, the ratio of the solvent and the SAS, their composition).

Regardless of the highr magnetisation of the metals, produced by this method, the magnetic characteristics of the magnetic fluids are not high. The low saturation magnetisation is explained by the large thickness of the SAS layer required for preventing bonding of the metal particles with high magnetisation. Increasing thickness of the stabiliser decreases the volume fraction of the magnetic and this results in a decrease of the saturation magnetisation. In addition, the metal particles are usually not spherical and this increases the forces of attraction of the particles (in comparison with the attraction of the spherical particles), i.e., it is necessary to increase further the thickness of the SAS layer. The use of insufficiently large SAS molecules for stabilising the particles may result in the aggregation of the particles and delamination of the fluid during long-term storage. These magnetic fluids with limited stability can be used in systems with constant mixing.

Another method of condensation – *electrolytic condensation* from aqueous solutions of salts of the metals of the particles which rapidly disperse in the liquid carrier in the presence of the stabiliser. The process is conducted in a two-layer electrolytic bath with a rotating cathode where the lower layer is the electrolyte solution, the upper layer is the solution of the stabiliser in the liquid carrier. The cathode rotates at a high speed so that the metal particles, which just managed to condense from the electrolyte on its surface, immediately penetrate into the dispersed medium and are coated with the SAS layer.

The electrolytic method can be used to produce metal particles of different dispersion depending on the spinning speed of the cathode, the depth of its immersion, the electrolyte concentrations, the temperature conditions, the cathode material. For example, the electrolytic method has been used to produce a magnetic fluid [11] with cobalt particles with the saturation magnetisation of 10 kA/m, the size of the particles was 30–80 nm. The small dispersion of the particles complicates the stabilisation of the magnetic fluid and restricts the concentration of the magnetic. The productivity of the method is low. Because of these shortcomings, the electrolytic method is not used widely.

High-dispersion powders can also be produced by *vacuum condensation* of the vapours of metals heated to high temperatures. The condensation process takes place mostly on the surface of the vessel walls in which the vacuum is produced, and the conditions of interaction of the metal atoms with the surface play the controlling role in the formation of colloidal particles. However, it is quite efficient to produce high-dispersion particles of metals by condensation of their vapour is directly in the disperse medium. Of considerable importance is the wetting of the surface of the resultant colloidal particles, i.e., their lyophilic nature in relation to the liquid carrier. In the presence of the lyophilic surface the particles are rapidly wetted by the disperse medium and their growth is interrupted. This method can be used to produce very small particles.

The vacuum condensation methods have not as yet been used for producing magnetic fluids because they are quite complicated. However, the development of vacuum technology has made it possible to obtain satisfactory results by combining vacuum evaporation of metals with their condensation in the fluid. Small dimensions and high magnetic properties of the particles, produced by vacuum evaporation result in high magnetic properties of the fluids. The specimens remain stable in the gravitational and magnetic fields, and x-ray diffraction analysis confirms the absence of metal oxides in the fluids (with the exception of a small amount of FeO). These data confirm the promising nature of the method of producing magnetic fluids with metals in in the form of magnetics, although in the currently available form the method is characterised by low productivity and is not suitable for large-scale production of magnetic fluids.

Currently, the most widely used condensation method is the chemical condensation of high dispersion magnetics, based on the reaction:

$$FeCl_2 \cdot 4H_2O + 2FeCl_3 \cdot 6H_2O + 8NaOH \rightarrow Fe_3O_4 \downarrow + 8NaCl + 20H_2O.$$

The method of chemical deposition of the high-dispersion magnetite is based on the rapid neutralisation during heating and constant mixing of the salts of bivalent and trivalent iron with an excess of the water solution of ammonia. The deposit formed during the reaction consists of magnetic particles with the size of 2–20 nm, with the average size of approximately 7 nm. The magnetic properties of the particles are close to those of the magnetite single crystals; the particle surface is characterised by high adsorption capacity which is important for stabilising the particles.

The method has considerable advantages in comparison with other methods of producing high dispersion particles of the magnetite, examined previously. In particular, it is characterised by high productivity – its productivity is limited only by the capacity of the chemical reactor because the rate of the actual reaction is very high. In addition, the method is suitable for industrial production and can be easily automated or mechanised.

13.2. The method of producing magnetite and magnetic fluids on transformer oil

At the present time, the chemical condensation method is the principal method of producing magnetic fluids. To ensure efficient synthesis of magnetite, it is necessary to adhere strictly to the proportions of the reagents and the method of working with them. The main technological operations of the method will now be described.

The initial reagents for producing the nanodispersed magnetite are the aqueous solutions of iron: $FeCl_3 \cdot 6H_2O$ and $FeSO_4 \cdot 7H_2O$. When using 25 g of magnetite it is necessary to use 138 ml of the $FeCl_3 \cdot 6H_2O$ solution, prepared by dissolving 0.8 kg of salt in a litre of distilled water. The solution can be used for one month from the day of preparation. To prepare the required amount of the $FeSO_4 \cdot 7H_2O$ solution, it is necessary to dissolve 38 g of salt completely in 75 ml of water with constant mixing. During dissolution, the solution must be heated in order to avoid the transfer of bivalent iron to the trivalent state. The solutions of the ferrous salts, produced by this method, are poured into a common vessel and subsequently 180 ml of 25% aqueous solution of NH_4OH is added with constant stirring.

The magnetite is produced by chemical condensation by the reaction:

$$2FeCl_3 \cdot 6H_2O + FeSO_4 \cdot 7H_2O + 8NH_4OH = Fe_3O_4 \downarrow + 6NH_4Cl + (NH_4)_2SO_4 + 4H_2O,$$

i.e., the co-precipitation of the salts of the bivalent and trivalent iron ($FeCl_3 \cdot 6H_2O$, $FeSO_4 \cdot 7H_2O$) with the 25% excess of the aqueous solution of ammonium hydroxide (NH_4OH). One of the conditions for producing the magnetite of the colloidal range of dispersion is the efficient stirring of the solutions in order to restrict particle growth.

To separate the magnetite, the vessel with the aqueous suspension of the magnetite and the reaction products is transferred onto a magnet where the magnetite is separated by magnetic separation from the mother liquor (reaction products). The aqueous suspension of the magnetite in the form of a deposit is at the bottom so that the mother liquor can be poured away by decanting. (Decanting in chemical laboratory practice and chemical technology is the mechanical separation of the solid phase of the disperse system (suspension) from the liquid phase by pouring away the solution above the deposit). The process of rinsing the suspension must be repeated several times. After the last rinsing, the vessel is again placed on the magnet in order to concentrate the aqueous suspension of the magnetite and remove the maximum amount of water from it.

The peptisation of magnetite, prepared by co-precipitation of the souls, is the next important stage of preparation of the magnetic fluid. Peptisation is the stabilisation of magnetite to produce the concentrate of the magnetic fluid in the aqueous medium. By heating up to 50–60°C a mixture of oleic acid in the transformer oil with the ratio of the components 1:1 (10 ml of transformer oil and 10 ml of oleic acid) is prepared. Subsequently, with constant manual mixing and heating on an electric plate, a mixture of oleic acid in transformer oil is added to the aqueous suspension of magnetite at 75–80°C. The process lasts 3–5 min.

To ensure that the magnetic fluid is stable and does not separate with time, the concentrate of the magnetic fluid must be dehydrated (paste). Therefore, the paste is rinsed at 75–80°C with ethyl alcohol until the alcohol becomes transparent (approximately 150 ml per 25 g of magnetite). The process is realised with the efficient manual mixing, with preheating the mixture of the paste and ethyl alcohol to 60–70°C. Final dehydration of the paste takes place in this case.

To disperse the concentrate of the magnetic fluid, small portions of transformer oil (~20 ml) are added to it. Preliminary mixing is carried out manually (using a rod with a fluoroplastic tip) and then the vessel with the magnetic fluid is placed on a heating ring and dispersion continues with mixing with an automatic stirrer for 1–2h at 80–85°C. The resultant magnetic fluid is homogeneous, black with a characteristic shine. Subsequently, the fluid is cooled to room temperature with continues mixing using the automatic mixer.

The preparation of the magnetic fluid is completed by centrifuging. For this purpose, the magnetic fluid is poured into plastic vessels and centrifuged with the separation factor of 6000 g for 1 h. After completing the process measurements are taken to determine its parameters (saturation magnetisation, density, viscosity, etc).

13.3. Laboratory equipment for producing magnetic fluids by chemical condensation

The described laboratory equipment was developed at the Belarusian National Technical University [11]. The equipment consists of a magnetically controlled reactor with a capacity of 3.5–4 l made of stainless steel (glass can also be used). The reactor consists of a cylindrical casing, a cover and a blade-type stirrer, Fig. 13.5.

Nozzles are installed in the cover of the reactor and used for supplying a mixture of 20% solutions of ferrous (II) and ferric (III) salts and 25% of the aqueous solution of ammonia.

A jacket is used for maintaining the thermal stability conditions of the process, with the heat carrier pumped through the jacket of the thermostat. The temperature is controlled with a chromel–copel thermocouple, installed in the cover of the reactor.

The components required for magnetic fluids are supplied in the process through three nozzles 7. A blade-type mixer 3 is secured to the shaft of the reactor for efficient stirring. At the bottom of the reactor there is valve 6 through which the completed magnetic fluid is discharged. A permanent magnet, positioned below the bottom of the reactor, is used for separating the aqueous suspension of the magnetite and magnetite and water.

The main parameters of the process of producing the magnetic fluid in this laboratory equipment are:

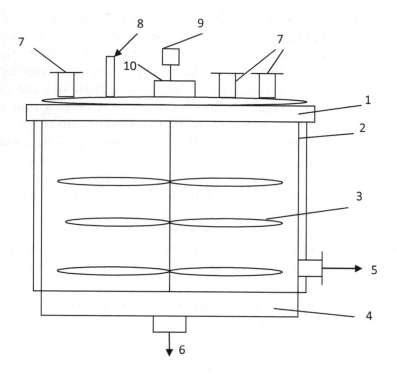

Fig. 13.5. The reactor for producing magnetic fluids. 1 – magnetically controller reactor; 2 – jacket; 3 – propeller mixer; 4 – electromagnet or permanent magnet; 5 – nozzle for discharge of rinsing water; 6 – nozzle for discharge of magnetic fluid; 7 – nozzles for pouring in starting reagents and acetone; 8 – thermocouple; 9 – power source; 10 – regulator of speed of rotation.

1. The concentration of the solutions of ferrous (II) and ferric (III) salts 20%;

2. The temperature of the process of producing magnetite 45–50°C;

3. The duration of magnetic separation of magnetite on the magnet 2–3 min;

4. The pH of the medium in which the peptisation of the magnetite deposit is carried out 12–14;

5. Peptisation temperature 50–55°C;

6. Peptisation time 2–3 min;

7. Dispersion temperature 70–80°C;

8. Dispersion time 1–1.5 h;

9. Cooling the magnetic fluid to a temperature of 25–30°C;

10. Centrifuging of the magnetic fluid with the separation factor of 6000 g for 1–2 hours resulting in the completed product.

The system is designed for synthesis of magnetic fluids based on kerosene, petroleum oils, polyethylsiloxane fluids PES-1,2,3,4 and perfluoro organic. The system is universal and efficient in service, its productivity is 0.2 l/3 h.

13.4. Selection of the dispersion medium

The selection of the dispersion medium used for the preparation of the magnetic fluid is dictated by its purpose and the type pf application. Depending on the problem to be solved, there are magnetic fluids based on water or a water-soluble base, and also based on hydrocarbons, organic silicon or organofluoric compounds. The formation of a stable colloid in each case has special features associated with the selection of the stabiliser, optimisation of the ratio of the ferrophase, the stabiliser and the base, and the transfer of the disperse ferrophase from one type of the medium to another [11].

Light hydrocarbons
The first magnetic liquid was produced in a ball mill by milling of magnetite in the presence of kerosene and oleic acid $C_{17}H_{33}COOH$. Since that time the magnetic fluids based on kerosene have become a classic object for investigations and production.

At present, the magnetic fluids based on kerosene are produced mostly by chemical condensation. The paste from which the water has been removed is a concentrate of the magnetic fluid and its saturation magnetisation is approximately 200 kA/m. The dilution of the fluid by the disperse medium produces a magnetic fluid.

It is important to know the problems formed in producing the paste and diluting it. Firstly, the paste must be dehydrated to prevent the formation of an emulsion which has a negative effect on the stability of the magnetic fluid. To ensure dehydration, the paste is heated during constant stirring, processed in filter presses and centrifuges. The removal of water from the paste is an operation which would ensure the stability of the magnetic fluid if performed with high efficiency.

The second operation important for producing the magnetic fluid is the dilution of the paste by the liquid carrier. Since the stabiliser is soluble in the disperse medium, then dilution may result in partial desorption of the molecules of the stabiliser from the surface of the particles and, consequently, agglomeration of the particles resulting in the growth of the size of the particles and, in the final analysis, in the delamination the magnetic fluid. To prevent this, dilution is carried out in a solution of a stabiliser in the liquid carrier resulting in the compensation of the desorption of the stabiliser from the surface of the particles and their complete coating with the layer of the surfactant molecules. The paste is diluted in heating to 60–70°C and constant stirring for several hours. Consequently, it is possible to separate using the moving rings of the long-chain molecules of the stabiliser the magnetite particles which merged in the deposit and the effect of magnetic forces into clusters. Therefore, a stable colloid system, consisting of the individual particles of magnetite, coated with the oleic acid layer, and the liquid carrier is produced.

In the laboratory conditions, to increase the stability of the system, the largest particles are removed from the system by centrifuging.

The magnetic fluids, produced by this method with magnetite as the disperse base are characterised by the magnetisation of up to 100 kA/m a high stability in gravitational and magnetic fields.

Petroleum oils
In a number of technical tasks is it necessary to use magnetic fluids based on oils – transformer, condenser, industrial, turbine, vacuum, etc. The production of a fluid with magnetite particles does not differ in this case from the production of fluids based on light hydrocarbons: magnetite is produced by the deposition method and this is followed by peptisation of magnetite by the solution of oleic acid in oil. However, because of the high viscosity of oils the removal of water from oils is associated with difficulties and the remaining water may prevent the formation of a stable colloid during dilution of the paste. Therefore, the paste is dehydrated by treatment with polar solvents, for example, acetone or ethyl alcohol. This procedure produces magnetite fluids bases on viscous oils such as turbine or vacuum.

With increasing viscosity of the base the resultant saturation magnetisation decreases and for the magnetic fluids based on viscous oils reaches 40–50 kA/m.

The application of magnetic fluids in the lubrication systems and sealers requires that they are efficient at elevated temperatures but the oleic acid used in most cases as a stabiliser is oxidised in this case and this may result in coalescence of the colloid. Therefore, work is being carried out to develop stabilisers with a low thermal oxidation capacity and also develop magnetic fluids with anti-oxidation additions.

The process of production of the paste from high-viscosity bases is associated with difficulties and requires efficient control of removal of water. If the water is not completely removed, dilution may be accompanied by the formation of a gel instead of the formation of a stable colloidal solution. Therefore, to obtain magnetic fluids from high-viscosity bases, and also from the bases in which the solubility of the stabilisation is very low, good results can be obtained using the method of replacing the base. This method is based on the fact that the magnetic fluid, produced from a low-viscosity base which efficiently dissolves the stabiliser, receives a polar agent causing flocculation of the particles with the stabiliser adsorbed on it. Subsequently, the particles are separated from the liquid carrier and repeated peptisation is carried out in another base.

This method can be used to produce magnetic fluids on the basis of which it is difficult to produce them by another method because of either the high viscosity or the volatility or instability. However, the replacement is possible only when using the base of the same type, for example, hydrocarbons.

Organic silicon (silicone) fluids
In many problems it is important to ensure that the working fluid has a low pressure of saturated vapours, is capable of working in a wide temperature range and in contacts with corrosive media. These requirements are satisfied to a large extent by organic silicon fluids in the form of polymer compounds whose molecules consist of alternating atoms of silicon and oxygen with the attached hydrocarbon radicals in the free bonds of silicon.

The magnetic fluids based on organic silicon are produced by different methods, for example, by dissociation of iron pentacarbonyl. These fluids are characterised by the saturation magnetisation of approximately 100 kA/m, and the size of the carbonyl iron particles reaches 20 nm. The latter circumstance results in the instability of the fluid during long time storage – the iron particles of these dimensions are characterised by high magnetisation and cannot be efficiently stabilised. However, in the seals in which the fluid is constantly mixed as a result of wobbling and vibrations, these fluids can be used efficiently.

Since the organic silicon fluids are mutually soluble with hydrocarbons, they can be produced efficiently by the methods of replacing the base.

Organic fluorine compounds
The organic fluorine compounds, in particular, perforated fluids, are characterised by a number of unique properties, so that they can be used in many applications in seals. They do not mix with water or water-soluble fluids or with oils so that the magnetic fluids based on them can be used for sealing the separators of liquid media. In addition to this, the organic fluorine compounds are almost completely chemically inert, i.e., they can be used for sealing chemically corrosive media.

The stabiliser for the production of organic fluorine magnetic fluids should also be of the organic fluorine nature because the oil-soluble stabilisers of the type oleic acid i.e. are insoluble in perforated fluids.

Water base
Water is a unique fluid with a number of anomalous properties and, therefore, the magnetic fluid based on water is an interesting object useful in varies applications, such as medicine, thermal engineering, etc.

A special feature of producing magnetic fluid based on water is the application of water-soluble surfactants. Another, more important special feature is that in the water which is a polar solvent, the magnetic particles may undergo hydration so that it is necessary to take special measures to protect solid particles against oxidation.

The simplest method of producing the magnetic fluid based on water is the one using the magnetite ferrophase. The colloidal particles of magnetite are produced by, for example, deposition of iron salts. The colloid of the metal oxides in the water medium can be stabilised using soaps of fatty acids, sulphonates, high-atomic alcohols, and esters. Although the magnetic fluids have been produced using almost all stabilisers, in the majority of cases their magnetic properties remained extremely low. The highly magnetic stable colloids can be produced using sodium laurisulphate and sodium oleate.

When producing the water-based magnetic fluids it is very important to maintain the optimum ratio of the magnetite and stabiliser because the deviations from the optimum ratio result in a large decrease of either the magnetic properties of the fluid or its stability. Nevertheless, regardless of these difficulties, researchers in a number of scientific centres have produced samples with high stability which do not change their magnetic properties after centrifugal and long-term storage.

In the water-based magnetic fluids the particles have a double layer of the stabiliser and, therefore, the saturation magnetisation of these fluids is lower than for the specimens based on other carriers. Nevertheless, regardless of the low magnetic properties, these fluids are used widely in areas where it is necessary to prevent any corrosive effect on the object, for example, the living organism.

13.5. Production of magnetic fluids with microdroplet aggregates

The magnetic fluids with the microdroplet aggregates belong in the group of highly magnetically sensitive fluids. They are produced by diluting the concentrated magnetic fluid – magnetite in kerosene with the solutions of oleic acid in kerosene of different concentration [57].

The colloidal solution contains microdroplet aggregates, if the starting fluid contains magnetite with a volume content of solid articles of 7–12% and is diluted with a 4–7% solution of oleic acid in kerosene to the following ratio of the components, wt.%: magnetite 2–3; SAS 1–2; kerosene 5–6; balance is 4–7% solution of oleic acid in kerosene. This produces a microemulsion containing two liquid phases – concentrated (microdroplet aggregates) and low concentration (initial fluid, diluted with the solution of oleic acid in kerosene to the concentration of the solid phase not higher than 1.3%).

When using higher concentrations of the magnetic fluids with the solid phase from 15 to 20% as the initial fluid for producing microemulsions with the optimum parameters as regards the visualisation of the defects, the diluent is represented either by pure kerosene or a solution of oleic acid in kerosene with the concentration at higher than 3%.

Questions for chapter 13

• *Which components must be included in the composition of the magnetic fluid? Why are they selected?*
• *What is the sedimentation stability of the magnetic disperse system?*
• *What is the main difference between the magnetic fluids with polar and non-polar dispersed media? Describe examples of such disperse media.*

- *Describe the method of crushing used for producing the magnetic fluid. What are the advantages and shortcomings?*
- *Characterise the condensation methods used for producing the magnetic fluid.*
- *Write the equation of the chemical reaction for producing the high-dispersion magnetite for the solution of ferrous salts.*
- *Describe the method and experimental equipment for producing the magnetic fluid by chemical condensation.*
- *What are the difficulties in producing the water-based magnetic fluid?*
- *Describe the applications of magnetic fluids with organic silicon or organic fluorine compounds as the disperse medium?*
- *Describe the method of preparation of magnetic microemulsions.*

14

Applications of nano- and microdispersed media

14.1. Applications of ferrosuspensions

The main practical application of ferrosuspensions (FS) is based on the so-called *magnetorheological* effect – the very strong dependence of viscosity on the strength of the magnetic field. This method is used for the development of magnetorheological and vibration dampers, magnetically controlled lubrication in friction sections and supports, sealing of threaded joints.

The possibility of processing components made of glass using magnetorheological suspensions (MRS) with abrasive particles was predicted in the middle of the previous century. The process of polishing using the magnetorheological suspensions for lapping was applied in 1980. In the year 2000, the company QED Technologies presented a method of polishing optical components of different shape and developed compositions of the magnetorheological fluids based on carbonyl iron. The shortcomings of the magnetorheological fluid developed by QED technologies include the relatively low rate of removal of the material, the short service life of the fluid and high cost.

In addition to this, the FS and paste-like compositions are used for visualisation of domain boundaries and in magnetic defectoscopy, in the manufacture of tape recorder tapes, the method of separation of iron ores and some other areas.

The detailed solution of all these problems has made it possible to determine the mechanical aspects of the polishing of different optical materials and complete the development of the composition of magnetorheological fluids which can be used for forming the surfaces with the surface irregularities of 2–4 atomic layers. These parameters fully satisfy the requirements of the optical industry in near future.

The application of the ferrosuspensions for damping vibrations of different devices and systems is also a method which has been developed to a large extent. The procedure may be described as follows. The magnetorheological vibration damper contains a casing filled with a damping medium with an electromagnetic system placed in it with a moving mass secured by a flexible bond with the casing. The use of the magnetorheological suspension as the damping medium regulates the frequency characteristics of the magnetorheological damper changing the voltage at the coils of the electromagnetic system.

The ferromagnetic suspensions are used for the development of sensors of the angle of inclination, acceleration, displacement, small pressure gradients, and also for the manufacture of flowrate metres and audio loudspeakers; they are also used for biomagnetic separation and bioresonance diagnostics in medicine.

The deflectors of the cosmic main parts use air dust shielding valves with permanent magnets. The valve consists of a rubber sleeve with clamps made of permanent magnets situated in two rows in the zigzag order, and protects the space below the deflector against the penetration of dust both on the starting platform and during the flight, reacting to excess pressure and inside wind. The deflector contains four valves with ferrosuspensions.

The disperse particles of the magnetorheological fluid form stable chain-like clusters under the effect of the magnetic field and result in the effect of the controlled fluidity and form.

Although the self-assembling in two measurements (2D) has been investigated sufficiently (especially using solid surfaces as substrates), the self-assembling in three measurements (3D) is far more complicated. The process of 3-D self-assembling of the diamagnetic plastic objects, maintained in the paramagnetic fluid by a heterogeneous magnetic field, has been described [52]. At the present time, this method is used on the scale from millimetres to centimetres but it may also be possible to use it in the range from ~10 mm to 1 m. The magnetic field and its gradient levitate the objects, organise their self-assembling of the substrate and influence the shape of the assembled clusters. The structure of the assembled 3-D objects can be made more complicated by the use of rigid mechanical substrates: either the walls of the container or co-levitating components which are spatially combined with the soft substrate by the gradient of the magnetic field. The mechanical effect makes the clusters stable and the addition of an adhesive substance (adhesive) followed by ultraviolet irradiation may result in joining of the components.

The results of laboratory investigations of the rheological properties of the magnetorheological fluids, designed for use in shock absorbers and vibration dampers have been generalised. The rheological properties of the fluids are described by the model of the viscous–ductile substances proposed by Herschel and Bulkley. The aim of the experiments was to determine the shear stress, the yield limit, the fluidity factor and the exponent in dependence on the density of the magnetic flux, followed by comparative examination of the rheological properties of the investigated fluids.

The Ferrari company uses magnetorheological fluids in some car models for improving the properties of the suspension. Under the effect of the electromagnet, controlled by a computer, the suspension can become instantaneously either more rigid or softer.

14.2. Using nanodispersed magnetic fluids in science and technology

The main now traditional applications of magnetic fluids will be listed.

Magnetic fluid seals (MFS)

The operating principle of the magnetic fluid seals is that a heterogeneous magnetic field, sustained in the gap by the magnetic fluid, is formed in the gap between the casing and the rotating shaft of the shaft carrying out reciprocal movement. This means that magnetic fluid overlaps the gap ensuring the required leak tightness with almost free movement of the shaft.

Cleaning water surfaces to remove oil per in failure spillage and catastrophes

The magnetic fluids based on kerosene are dissolved in oil products transforming them to low-concentration magnetic fluids. The resultant solutions are collected by a magnetic 'trap' and subsequently the oil products can be separated from the magnetic fluid using a heterogeneous magnetic field.

Cooling fluids in machining of metals and cooling of transformers
Thermomagnetic convection may be more efficient than natural convection and, therefore, the application of the magnetic fluids as heat carriers is highly efficient in devices with a strong magnetic field. In particular, the replacement in the transformers of the conventional transformer oil by the magnetic fluid, prepared on the basis of the same oil, may greatly increase the permissible power.

Fine separation of non-magnetic materials with different density
The effect of controlled magnetic levitation on the magnetic fluid is used for the processes of separation of the required fraction from a multicomponent disperse system. This method is economically most efficient in separation of non-magnetic particles of non-ferrous (expensive) metals and minerals.

Sensors of the level and slope
The magnetic fluids are characterised by two 'mutually excluding' properties – fluidity and the relatively high magnetic susceptibility. Therefore, the application of the magnetic fluid as active elements in the sensors of the level and angle of inclination (slope) makes it possible, firstly, to use a simple and reliable method of electromagnetic indication and, secondly, greatly improves the accuracy of measurements.

Acoustic wide-band dynamics of higher power
The interpolar gap of the acoustic dynamics with a strong heterogeneous magnetic field is filled with a magnetic fluid and, consequently, the main functional elements of the dynamics – the inductance coil, actually 'floats' in the fluid. In this case, the magnetic fluid plays the role of the heat-conducting element and of the acoustic contact lubricant thus greatly widening the dynamic range of the equipment and its amplitude–frequency characteristic.

Magnetically controlled transport of drugs
Using the magnetic fluid controlled by the magnetic field, it is possible to supply to the required area of the organism, for example, the area with a malignant tumour, medical drugs disperser in the fluid. The effect of the drug can be localised and can be maintained in the given area for a very long time. However, there is still a problem of producing the optimum (from the viewpoint of side

effects) composition of the magnetic fluid and the complete removal of the magnetic nanoparticles from the human organism after completing the healing process.

Magnetic flaw inspection and visualisation of videosignal recordings

Depositing a thin layer of a magnetic suspension or a colloid with microdroplets on the magnetised surface, it is possible to visualise, using an optical microscope, patterns produced as a result of the concentration of microparticles (microdroplets) in the areas of the specimen in which the sign of the magnetisation changes. This phenomenon is the basis of magnetic flaw inspection which makes it possible to reveal defects such as non-metallic and slag inclusions, cavities in components, delamination, weld defects, cracks, and also magnetic recording defects. In contrast to the dry powders of the ferromagnetics, the microdroplets are capable of moving at a high speed to the areas with defects because, being in the suspended state, they are not subject to any friction

Magnetic fluid – lubricant

The magnetic field can be used to sustain the magnetic fluid, playing the role of a lubricant, in the contact zone of the rubbing surfaces. In this case, the consumption of the lubricant and the number of stoppages of the mechanism are greatly reduced. The magnetic lubricant is based on organic silicon fluids and mineral oils are characterised by high anti-seizure properties and can be used at temperatures up to 300°C.

In the last couple of decades, magnetic fluids have also been used in scientific studies.

Acoustomagnetic identification of the soundwave

V.M. Polunin and I.E. Dmitriev [51, 59] investigated in 1997 the dispersion of the speed of sound in a fluid–cylindrical shell system by the method based on the application of the acoustomagnetic effect in a magnetic fluid. In the acoustomagnetic effect, the propagation of the sound wave in the magnetised magnetic fluid is accompanied by a disruption of magnetisation leading to the formation of a variable EMF in the inductance coil adjacent to the boundary of the sound beam.

Instead of the 'conventional' fluid the shell is filled with a magnetic fluid, and the role of the receiver of sound oscillations

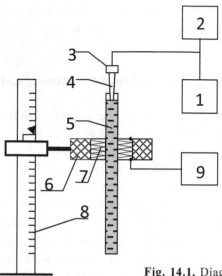

Fig. 14.1. Diagram of experimental equipment.

is played by a magnetic head placed outside the limits of the shell. Consequently, it is possible to move the receiver along the pipe with the fluid over large distances without disrupting the acoustic field in the fluid. It is not necessary to take into account diffraction disturbances, introduced by the hydrophone.

The magnetic head consists of an inductance coil and a source of a direct magnetic field.

Figure 14.1 shows the diagram of experimental equipment. The source of the elastic oscillations is the emitter 3, containing a piezosheet with a resonance frequency of 1 MHz. Measurements are taken at frequencies not higher than 200 kHz, i.e., away from the resonance frequency and, therefore, its frequency characteristic again be regarded as linear (having a plateau form). The source of the alternating EMF is the oscillator 1. The elastic waves are introduced into the magnetic fluid 5, filling the tube, through the waveguide 4. The magnetic head includes the permanent ring-shaped magnet 6, magnetised along the axis, and the inductance coil 7, positioned inside and rigidly bonded with it. The coil and the pipe are separated by an air gap. The ring-shaped magnet in the area in which the inductance coil is situated magnetises the fluid preferentially along the axis of the pipe. The alternating EMF, induced in the coil during propagation of the sound wave in the fluid, travels to the input of the oscilloscope 9. The magnetic head is placed on the kinematic section of the cathetometer 8 and moves freely along the pipe.

As in the conventional ultrasound interferometer, the measurement of the speed of sound in the fluid is reduced to determining the number of standing waves N_s, distributed along the length ΔL. Calculations are carried out using the equation

$$c = 2\Delta L v/N_s.$$

The resultant experimental data for the dispersion of the speed of sound in the fluid–shell system using shells with a wide spectrum of the geometrical and elasticity parameters show the efficiency of using the acoustomagnetic effect for investigating systems of this type. The proposed experimental procedure can be used for measurements in systems of different type, including in those in which it is difficult to obtain theoretical results.

Magnetic fluid membrane – model of the magnetic sealer
At present, the magnetic fluids are used mostly in magnetic fluid seals (MFS). Of considerable importance are the strength and kinetic properties of these devices. However, the investigation of these devices is the most complicated experimental task. It is therefore interesting to investigate the magnetic fluid membrane (MFM) [3] which may be used as a model of the MFS. The main element of the MFM is the magnetic fluid membrane which has been described in detail in chapter 11. In the MFM the magnetic fluid membrane overlaps the cross-section of the pipe with the container attached to it and used as a model of the object to be sealed.

Now follows a list of the promising applications of nanodispersed magnetic fluids:
    1. The properties of the magnetic fluid are efficiently controlled by the external magnetic field so that this offers considerable possibilities for technical and biomedical applications. Investigations are being carried out into the possibilities of using magnetic fluids for producing composite systems with a magnetic fluid designed for directional supply of medical drugs.
    2. Tests are being carried out of the possibilities of using the magnetic fluid as an active carrier of the compounds for the antibacterial effect on *Staphylococcus aureus* and *Escherichia coli*.
    3. The levitation effects which can be easily realised in the magnetic fluid are used in the design of separators and density meters

of non-magnetic materials, high sensitivity triaxial acceleration metres, and a number of other advanced devices. The effect of the processes of magnetophoresis and Brownian diffusion on the levitation of solids in the magnetic fluid is being studied [53].

4. Studies have been published describing new devices with the operating principle based on magnetic levitation and magnetic fluids. For example, in one of the later studies, it is proposed to use a magnetic fluid drive whose operating principle is based on the levitation of a non-magnetic solid immersed in a magnetic fluid in a magnetic field.

5. The US Air Force and Army have introduced a radio-absorbing coating based on a ferromagnetic fluid. Reducing the reflection of the electromagnetic waves, these coatings decrease the effective area of scattering of the aircraft.

6. The magnetic liquids have a large number of applications in optics because of their refractive properties. It is important to mention the measurement of the specific viscosity of the fluid, placed between a polariser and an analyser, illuminated with a helium–neon laser.

7. As a result of the application of the surfactants and a large number of particles carrying electrical charges in the magnetic fluids, they represent effective foaming additions. The addition of the magnetic fluid based on kerosene or water to the fuel for internal combustion engine in the volume ratio of 1:350 per quarter decreases the content of nitrogen oxides in the exhaust gases and increases the combustion efficiency of fuel.

8. If the magnetic fluid with different susceptibility (for example, as a result of a temperature gradient) is subjected to the effect of the magnetic field, this results in the formation of a heterogeneous magnetic volume force leading to a mechanism of heat transfer referred to as thermomagnetic convection. This form of heat transfer may be used in applications where it is not possible to use commercial convection, for example, in microdevices or in conditions with reduced gravitation.

9. The effect of the magnetic dependence of the transparency of the fluid can be used for deep low-frequency modulation of light. This principle is used for the dynamo-magneto-optical sensors of the shear strain rate. The sensors of this type to the magnitude and not rate of strain. The sensitivity of these sensors is sufficient for producing microphones and sensor devices of the type of displays

of laptop computers with the sensor introduction of commands and and data.

10. Heating the magnetic fluid above the Curie point T_C, we can greatly decrease its magnetic susceptibility. This is the basis of thermomagnetic convection. The layers of the magnetic fluid with $T < T_C$ are characterised by high magnetic susceptibility and are 'pulled' into the regions with higher strength of the magnetic field, displacing the layers with $T > T_C$. The intensity of thermomagnetic convection may be many times greater than that of gravitational convection.

11. Recently, reports have been published showing that the replacement of spherical nanoparticles in the 'conventional' magnetic fluid by magnetic nanotubes results in a stronger dependence of viscosity on the magnetic field thus widening the possibilities of the service application of the magnetoviscous effect [54].

12. The data obtained in examining the processes of capture, sealing and fragmentation of the gas cavity by the magnetic fluid, controlled by the magnetic field, can be used in the development of systems for taking gas samples, storage and subsequent analysis, counters and the devices for measuring the gas flow rate, used in chemical engineering, pharmaceutics [55, 56].

High-quality monitoring of the process of metered supply of small portions of the gas can be carried out using the results of experimental studies of the phenomenon of separation of the air bubble from the cavity in the magnetic fluid compressed by the ponderomotive forces of the magnetic field [58]. The bubbles, separated from the cavity, carry out pulsating motion in the magnetised magnetic fluid. The process is accompanied by electromagnetic radiation. The spectrum of this radiation contains ν – the frequency of oscillations of the bubbles [58, 59]. On the other hand, the frequency of oscillations of the bubble with a radius R_0 in the fluid can be determined from the equation [45]:

$$\nu = \left(2\pi R_0\right)^{-1} \sqrt{3\gamma P_0 / \rho},$$

where P_0 is hydrostatic pressure, $\gamma = C_p/C_V$ is the ratio of the specific heats of the gas, ρ is the density of the sample of the magnetic fluid.

Consequently, the radius of the bubble can be calculated from the equation:

$$R_0 = (2\pi v)^{-1} \sqrt{3\gamma P_0 / \rho}.$$

In the investigated example ρ = 1320 kg/m³, γ = 1.4, P_0 = 10⁵ Pa, the experimental value v = 2.5 kHz and, correspondingly: R_0 = 1.14 mm. Calculating the mass of the gas in the bubble using the equation: $\Delta m = 4\rho_g \pi R_0^3/3$ and assuming ρ_g = 1.29 kg/m³, we obtain $\Delta m \approx 8 \cdot 10^{-9}$ kg.

The results show that it will be possible to develop a new method of metered supply of small portions of gas into a reactor.

Questions for chapter 14

- *Name the applications of magnetorheological suspensions.*
- *Where can the magnetic levitation effect be used?*
- *Name the traditional areas of application of magnetic fluids.*
- *Name the restrictions for the use of magnetic fluids in medicine.*
- *What is the acoustomagnetic effect and how can it be used in practice?*
- *Where are the magnetic fluid seals used?*
- *Describe the application of magnetic fluids in optics.*
- *Describe the application of the process of capture, sealing and fragmentation of the gas cavity by the magnetic fluid controlled by the magnetic field.*

Conclusions

The nano- and microdispersed systems of the natural and artificial origin are used efficiently in processes. Of special interest is a group of fluid magnetised media – nanodispersed magnetic fluids and ferromagnetic suspensions.

The textbook describes the problems associated with the mechanics of liquid nano- and microdispersed magnetised media. The equations of dynamics of fluid magnetised media and their applications are described. Attention is given to the physical phenomenon specific for nano- and microdispersed systems: interfacial heat exchange, magnetocaloric effect; slipping of nano- and microparticles in relation to the liquid matrix; the magnetisation hysteresis of microdispersed suspensions; superparamagnetism of nanodispersed magnetic fluids; the non-Newtonian nature of the rheology of suspensions; magnetic levitation; diffusion and magnetophoresis; magnetorheological effect. The effect of the dimensional factor on the physical properties of materials is demonstrated. The methods of producing nanodispersed magnetic fluids and ferromagnetic suspensions and also the main and promising directions of these media with the appropriate physical interpretation are described.

A special feature of this book is that it is based on the results of many years of experimental studies of magnetic disperse systems. The scientific school of Prof. V.M. Polunin is well-known in Russia and abroad, and this textbook publishes under his supervision the scientific articles converted for use in textbooks. At the same time, the book does not include a number of problems concerning, for example, the special features of propagation of soundwaves in magnetic colloids and magnetorheological media, magnetically controlled processes of capture and subsequent disintegration of the air cavity by the magnetic fluid. Therefore, the authors plan further work to expand the range of subjects suggested for study together with other disciplines.

Since the nanotechnologies represent the rapidly developing scientific and technical direction, this textbook is published at the right time. It enables to study the current physical–chemical models of the processes, in which the active elements are nanotechnological materials, characterised by the unique 'mutually excluding' properties (such as fluidity and magnetic control) and are highly promising for application in different areas of technology.

References

1. Shliomis M.I., Usp. Fiz. Nauk, 1974. V. 112. No. 3. 427–459.
2. Polunin V.M., Magnetic fluids, Great Russian Encyclopedia: V.18. Lomonosov-Manizer, Moscow, 2011, 373–374.
3. Polunin V.M., The acoustic properties of nanodispersed magnetic fluids. Moscow, Fizmatlit, 2012.
4. Berkovskii B.M., et al., Moscow, Khimiya, 1989.
5. Odenbach S., (Ed.). Colloidal Magnetic Fluids: Basics, Development and Application of Ferrofluids, Lect. Notes Phys., Berlin, Springer, 2000.
6. Rosensweig R.E., Ferrohydrodynamics, Cambridge Monographs on Mechanics and Applied Mathematics, New York, 1985.
7. Ultrasound. Small Encyclopedia, ed. I.P. Golyamina, Moscow, Soviet Encyclopedia, 1979.
8. Zolotukhin I.V., et al., New directions of materials science, textbook, ed. B.M. Darinskii, Voronezh, VSU, 2000.
9. Tarapov I.E., Magn. gidrodinamika, 1972, No. 1, 3–11.
10. Tarapov I.E., Prikl. Matem. Mekh., 1973, V. 37, No. 5, 813–821.
11. Bashtovoi V.G., et al., Introduction to the thermomechanics of magnetic fluids, Moscow, 1985.
12. Landau L.D., Lifshitz E.M., Theoretical physics. Hydrodynamics. Moscow, Nauka, 1988, V. 6.
13. Rytov S.M., et al., ZhETF, 1938, V. 8, No. 5, 614–626.
14. Vladimirskii V.V., Nauchn. Sb. Student. MGU, Ser. Fizika, 1939. V. 10. S. 5-30.
15. Chechernikov V.I., Magnetic measurements, Moscow, MSU, 1969.
16. Vonsovskii S.V., Magnetism, Moscow, Nauka, 1971.
17. Polunin V.M., Magn. gidrodinamika, 1985, No. 3, 53–56.
18. Polunin V.M., et al., Journal of Magnetism and Magnetic Materials, North-Holland, 1990, No. 85, 141–143.
19. Baev A.R., Author Cert. No. 713599 USSR. A method for generating acoustic oscillations, 02/22/78, publ. 1980, Bull. No. 5.
20. Polunin V.M., Magn. gidrodinamika, 1978, No. 1, 141–143.
21. Polunin V.M., Akust. Zh., 1978, V. 24, No. 1, 100–103.
22. Polunin V.M., ibid, 1982, V. 28, No. 4, 541–546.
23. Berkovskii B.M., et al., Magn. gidrodinamika, 1986, No. 1, 67–72.
24. Polunin V.M., et al., Bulletin KurskGTU, 2003, No. 2 (11), 29–34.
25. Orlov D.V., et al. Magnetic fluids in mechanical engineering, Moscow, Mashinostroenie, 1993.
26. Lobova O.V., et al., Izv. Kursk. Gos. Tekhn. Univ., 2002, No. 1 (8), 286–294.
27. Lobova O.V., et al., Vibrating machines and technologies, Proc. 5th Internat. conf.;

Kursk State. Tekh. Univ. Kursk, 2001, 356–359.
28. Rabinovich M.I., Trubetskov D.I., Introduction to the theory of oscillations and waves, Moscow, Regular and Chaotic Dynamics Research Centre, 2000.
29. Drozdova V.I., et al., Magn. gidrodinamika, 1981, No. 2, 17–23.
30. Bratukhin Yu.K., Lebedev A.V., ZhETF. 2002, V. 121, No. 6, 1298–1305.
31. Polunin V.M., et al., Magn. gidrodinamika, 2012, V. 48, No. 3, 557–566.
32. Polunin V.M., Magn. gidrodinamika, 1979, No. 3, 33–37.
33. Frenkel' Ya.I., Kinetic theory of liquids, Leningrad, Nauka, 1975.
34. Bibik E.E., The interaction of particles in ferrofluids, The physical properties and fluid dynamics of dispersed ferromagnets, Sverdlovsk, Urals Scientific Centre of the USSR Academy of Sciences, Ufa, 1977.
35. The investigation of the propagation of ultrasound in the magnetic fluid: Report No. 3236, Institute of Mechanics, Moscow State University, Gogosov V.V., Moscow, 1985. 84 p, No. 77066746.
36. Polunin V.M., Akust. Zh., 1985, V. 31, No. 2, 234–238.
37. Isakovich M.A., ZhETF, 1948, V. 18, No. 10, 907–912.
38. Isakovich M.A., Mandel'shtam L.I., Usp. Fiz. Nauk, 1979, V. 129, No. 3, 531–540.
39. Mikhailov I.G., et al., Fundamentals of molecular acoustics. Moscow, Nauka, 1964.
40. Rzhevkin S.N., The course of lectures on the theory of sound, Moscow, MSU, 1960.
41. Shliomis M.I., ZhETF, 1971, V. 61, No. 6 (12), 2411–2418.
42. Naletova V.A., Shkel' Yu.M., Magn. gidrodinamika, 1987, No. 4, 51–57.
43. Kashevskii B.E., et al., *ibid*, 1988, No.1, 35–40.
44. Landa P.S., Rudenko O.V., Akust. Zh., 1989, V. 35. No. 5, 855–862.
45. Sirotyuk M.G., Acoustic cavitation, V.I. Il'ichev Pacific Ocean Institute, Russian Academy of Sciences, Moscow, Nauka, 2008.
46. Mikhailov I.G., Polunin V.M., Akust. Zh., 1973. V. 19, No. 3, 462–463.
47. Rayleigh L., Philos. Mag., 1917, V. 34, 94–100.
48. Chikazumi S., Physics of Ferromagnetism, 2nd edition, Oxford University Press, 1997.
49. Vand V., J. Phys. Coll. Chem., 1948, V. 52, No. 2, 227–299.
50. Bibik E.E., Kolloidn. Zh., 1973, V. 35, No. 6, 1141–1142.
51. Dmitriev I.E., Polunin V.M., Acoustical Physics, 1997, V. 43, No. 3, 295–299.
52. Ilievski F., et al., Soft Mater., 2011, No. 7, 9113.
53. Bashtovoi V.G., et al., Magnetohdrodynamics, 2008, V. 44, No. 2, 121–126.
54. Zhenyu Wu, et al., Phys. Status Solidi B 247, No. 10, (2010), DOI 10.1002 / pssb.201046208, 2412–2423.
55. Polunin V.M., et al., 2012, V. 48, No. 3, 557–566.
56. Bashtovoi V.G., Nanotekhnika, 2013, No. 33, 84–90.
57. Shagrova G.V., Methods of control of information on magnetic media. Moscow, Fizmatlit, 2005.
58. Boev M.L., et al., Acoustical Physics, 2014, V. 60, No. 1, 29–33.
59. Polunin V.M., Acoustics of nanodispersed magnetic fluids, New York and London, CRC Press and CISP Cambridge, 2015.

Index

Milton Keynes UK
Ingram Content Group UK Ltd.
UKHW040058071024
449327UK00019B/633